BIOINFORMATICS LAW

Legal Issues for Computational Biology in the Post-Genome Era

Jorge L. Contreras and A. James Cuticchia, Editors

ABA Section of Science & Technology Law

Cover by Amanda Fry/ABA Publishing.

Printed in the United States of America

17 16 15 14 13 5 4 3 2 1

Cataloging-in-Publication Data is on file with the Library of Congress

Bioinformatics Law: Legal Issues for Computational Biology in the Post-Genome Era
Edited by: Jorge L. Contreras and A. James Cuticchia
ISBN: 978-1-62722-010-1

Discounts are available for books ordered in bulk. Special consideration is given to state bars, CLE programs, and other bar-related organizations. Inquire at Book Publishing, ABA Publishing, American Bar Association, 321 North Clark Street, Chicago, Illinois 60654-7598.

www.ShopABA.org

Summary Table of Contents

Contents

CHAPTER 15

Informatics and Public-Health Surveillance 191

Leslie P. Francis and John G. Francis

Foreword

In 2000, the leaders of the public Human Genome Project (HGP) and privately funded Celera Genomics, at a White House ceremony presided over by President Bill Clinton, announced that they had completed a "first draft" of the human genome. A year later, the competing groups published their groundbreaking results and methodology in *Science* and *Nature*. And by 2003, all agreed that the "reference" human genome sequence was essentially complete, bringing the world's largest scientific collaboration to an end.

A decade has passed since that history-making event. The human genome sequence is being studied in thousands of laboratories across the world, new associations between genetic variations and disease and physiological traits are announced each week, and the databases containing the accumulated genomic data of the research community are growing at rates that are orders of magnitude greater than those seen by the original HGP. We are living in what many have termed the "post-genome" era, a time in which the availability of the human genome sequence is taken as a given, and in which it serves not as the goal, but as a starting point for multiple new and innovative research endeavors.

One of us (Cuticchia) is a bioinformaticist who turned to law, and the other (Contreras) is a lawyer who came late to science. We have each been privileged to contribute in our own small ways to the amazing advances that have characterized genomic science and technology over the past two decades. Now we are active participants in the ongoing work of the post-genome era, and we continue to be awed and inspired by the breathtaking pace of change in this field.

From our respective vantage points, however, we have each been struck by the lack of resources and reference materials elucidating the intersection of bioinformatics and the law. We met while serving together on the National Conference of Lawyers and Scientists, a joint committee of the American Bar Association's Section of Science & Technology Law and the American Association for the Advancement of Science, and we soon realized that such a reference work would come into being only if we created it ourselves. This book is the result.

In this volume, we have collected cutting-edge research and commentary from leading scholars, bioethicists, and legal practitioners in an attempt to cover the broad and ever-changing intersection of law and bioinformatics. We begin with an overview of the evolving bioinformatics field (Chapter 1). We then explore legal issues surrounding the software tools that enable large-scale computational biology, including procurement and licensing of commercial software systems (Chapter 2), licensing of open source software (Chapter 3), and in-house and contracted software development (Chapter 4). We then move to the complex intellectual property landscape that characterizes the field, first addressing patents on computer software, algorithms, and related inventions (Chapter 5), and then moving to the highly contested issues surrounding patents on human genetic material (Chapters 6 and 7). We next turn to legal aspects of the research environments in which bioinformatics work is increasingly conducted, including biopharma collaborations (Chapter 8) and university research and development (Chapter 9). Chapter 10 addresses the novel legal issues arising from the development and use of standardized technologies in bioinformatics, and Chapters 11 and 12 address the protection and use of databases in bioinformatics research. The final three chapters cover increasingly important issues relating to patient and research subject data, including information security and privacy (Chapter 13), informed consent (Chapter 14), and public health surveillance (Chapter 15).

The field of bioinformatics continues to evolve rapidly, while the legal rules, doctrines, and practices attempt to catch up. We hope that this volume will be a useful resource for lawyers, researchers, policymakers, and scholars who seek insight into the complex intersection of law and bioinformatics in this post-genome era.

The Editors

Acknowledgments

The editors would like to thank the Book Publishing Board of the ABA Section of Science & Technology Law for their support, encouragement, and patience throughout this project, with special thanks to Sarah Forbes Orwig of ABA Publishing. In addition, this book would not have been possible without the hard work of Dean's Fellows Chris Pepe and Sydney Kestle at American University Washington College of Law.

CHAPTER 1

Bioinformatics Overview

A. James Cuticchia

INTRODUCTION

Let us begin with a riddle, "What has no beginning and has no definition?" The answer, of course, is *bioinformatics*. This most elusive of disciplines is the son of many fathers. To try to define bioinformatics is a fool's errand; and thus we begin with a riddle.

A Google search in 2012 for the word "bioinformatics" yields more than 34 million results. A search for "bioinformatics definition" yields over 4 million results.

The *Oxford English Dictionary* defines bioinformatics as "the science of collecting and analyzing complex biological data such as genetic codes." However, one author has proposed the following definition:

> Bioinformatics is conceptualizing biology in terms of molecules (in the sense of Physical Chemistry) and applying "informatics techniques" (derived from disciplines such as applied math, computer science and statistics) to understand and organize the information associated with these molecules, on a large scale. In short, bioinformatics is a management information system for molecular biology and has many practical applications.[1]

Thus, is bioinformatics applying *informatics to biology*? This is probably not the best way to define a word. *Black's Law Dictionary*, however, does not even provide a definition.

The purported first use of the term "bioinformatics" comes from Paulien Hogeweg, who wrote: "In the beginning of the 1970s, Ben Hesper and I started to use the term 'bioinformatics' for the research we wanted to do, defining it as "the study of informatics processes in biotic systems." Hogeweg also writes: "I received a surprising e-mail from Oxford University Press: 'It appears you be [sic] responsible for the term, bioinformatics.'"[2]

The supposition here is much the same as "Yes, Virginia, there is a Santa Claus. If you see it in *The Sun* it must be true." If Oxford University Press says you are responsible for a definition, it also must be true.

The purpose of these examples is not to attack any one definition. Rather, it is to give a peek into the difficulty of defining this topic. As this is a book on legal issues pertaining to bioinformatics, it is of note that the courts *have* attempted to define bioinformatics.

In *Jin Xiong Petitioner v. Commissioner of Internal Revenue*, the United States Tax Court stated that bioinformatics is "the analysis of information relating to biological structures with the aid of computers."[3] The case dealt with the deduction of expenses related to the writing of a book on bioinformatics. Thus, much like it took the IRS to bring down Al Capone, it takes the IRS to establish a legal definition of bioinformatics!

BIOLOGY

This section is a primer on biology—in particular, genetics—and is provided to put the databases and software used in bioinformatics in some context. Many attorneys' only exposure, if any, to genetics is in the area of DNA evidence.

While many people know the mantra "We are made up of DNA," that statement is both simplistic and incorrect. (We are made up of proteins!) For the most part, all cells contain DNA. DNA is tightly wrapped into structures called chromosomes. DNA is so densely packed that if unwound, the DNA in a single human cell would have a length of 3 km.[4] If one were to remove all of the DNA from each human cell, it would reach a length equal to 70 round-trips from the Earth to the Sun.[5] While we are not "made of DNA," we certainly have a lot of it!

DNA contains the blueprint of how to make a cell. The information is stored in a structure, which is composed of four different "bases." These bases are: adenine, thymine, cytosine, and guanine. Since DNA is a double helix, the helical nature (stair steps) is made by the pairing of two bases. Adenine pairs with thymine, while cytosine pairs with guanine. Because of this pairing, each strand of DNA is complementary to its sister strand.

For example a DNA sequence might read:

AAGTAGCCATCAGTGACT (Strand 1)
TTCATCGGTAGTCACTGA (Strand 2)

This relationship between the two strands makes copying of the molecule simple. During replication, the cell process of copying DNA, each strand will separate and new bases of DNA will be added to each strand:

AAGTAGCCATCAGTGACT (Strand 1)
TTCATCGGTAGTCACTGA (Strand 1 Copy)
TTCATCGGTAGTCACTGA (Strand 2)
AAGTAGCCATCAGTGACT (Strand 2 Copy)

Since all cells, with some minor exceptions, contain copies of the DNA molecule, they have the genetic information needed to make any type of cell. Thus, a cell in the liver has the same information as a muscle cell. While it may have the same information, other biological processes related to the structure and formation of that cell limit what it can produce. Stem cells are pluripotent cells, which have the ability to produce any type of cell. Since an embryo starts with a single cell, that cell must have the ability to make any other type of cell (liver, brain, muscle, etc.). Since the stem cell is undifferentiated, it can serve as a valuable model for scientific experiments.

The DNA in a cell is stored inside a membrane called the nuclear membrane, which defines the nucleus. This is where DNA is unwound for copying (replication) and for expression (transcription). If a cell is dividing, it is necessary that each cell contain a copy of the DNA. Therefore, the cellular "machinery" necessary for quickly making a copy of all DNA on all the chromosomes is contained in the nucleus.

In order for DNA to make a protein (e.g., hemoglobin), the DNA must be transported outside the nucleus into the cell's interior where the machinery for protein production exists. DNA is not used, directly, to transmit its information to the protein production machinery of the cell. Rather, a "working copy" of the DNA is created and exported from the nucleus. This process is called transcription.

The "working copy" of the DNA is made by copying one strand of the double-stranded helix with RNA. This molecule is identical in information but is structurally different from the DNA molecule. One example is that mRNA contains substitutes on the four DNA bases (thymine is substituted with uracil) and is single-stranded (thus representing only one of the sister strands).

Once the cell needs to make a particular protein, the DNA containing that information will be transcribed and exported outside the nucleus. It is at this point the gene is termed "expressed." Much of the area of bioinformatics is devoted to examining when and to what degree genes are expressed. By measuring mRNA concentrations, one may obtain an estimate on how actively a gene is being expressed under a particular condition.

The mRNA is fed into the protein production component of the cell, the ribosome. Proteins are made up of amino acids. These amino acids are assembled in a linear manner, and then the amino acid chain will "fold" into an active protein state. A genetic "word" is made up of three letters. Thus G-G-G would be the instruction to add a glycine amino acid to the protein chain.

The process by which DNA—> RNA—> PROTEIN is called the "Central Dogma" of molecular biology.

Some Examples of the Central Dogma at Work

Production of Insulin

When a person eats, insulin is produced in order to regulate the amount of sugar in the blood. Without insulin, such sugar would become toxic. Once sugar enters the bloodstream, it will reach the pancreas. The pancreas is the only organ that produces insulin in humans. Pancreatic B-cells will detect the presence of glucose on the outer membrane of the pancreas, and that will start the cellular machinery needed to produce the insulin.

First, chromosomes are unwound in order to make the insulin gene accessible for transcription. Signals to the nucleus start the transcription process. One strand of DNA is transcribed into mRNA. Since DNA is a double helix with complementary strands, it is important that the correct strand is transcribed. Discussion of this selection is well outside the scope of this book.

Once the mRNA is created, it is sent from the nucleus into the cell interior, called the cytoplasm. There, the mRNA attach to the protein production center, the ribosome. The mRNA is read, three letters at a time, and molecules containing RNA on one end and an amino acid on the other end connect with the letters to produce a strand of amino acids as the mRNA strand is read by the ribosome. The completed protein, insulin, is then exported outside the cell.

Replacing Skin

When a person suffers third-degree burns, the three layers of the skin (epidermis, dermis, and hypodermis) are harmed. The healing process requires that new skin cells be produced. Unlike in the insulin example where a cell is performing a particular internal function, here a cell is *creating* other complete cells. The process is quite different.

In general, skin cells are "programmed" to reproduce. Once the signal is given to create new cells, a cell grows and begins to replicate its DNA. Since the cell will be dividing into two, it is important that enough machinery is available to support two cells. While this is going on, the chromosomes begin to unwind.

Unlike in the insulin example, where only the insulin gene had to be accessible, here all DNA must be unwound. As that occurs, the strands of the DNA separate, creating two templates. Cellular machinery in the nucleus then begins to copy each of the strands. Once the copying is complete, two complete strands exist, each made of one strand of the existing DNA and its copy. As this process of replication is ongoing, the replicated strands begin again to package themselves as chromosomes. At its completion, each chromosome now has complete strands of DNA. The chromosomes align themselves in such a way that their splitting results in a full complement of DNA on each side of the nucleus. The nucleus then divides.

Each nucleus will be separated in such a way as to be allocated to a different cell during cell division. Once all the processes in the cell needed for division are completed, the cell will split into two cells, each having an identical set of DNA.

Cancer (Breast Cancer)

While there are numerous types of cancer, there is one underlying feature of them all—rapid, abnormal cell growth. The breast is made up of three cell types: (1) milk-producing cells; (2) milk ducts; and (3) fatty tissues. While the exact causes of breast cancer are an active part of cancer research, the focus in this example is what is happening on a cellular level.

A cell in the breast receives a signal to begin to divide. The cells "change" from normal cells into mutant cells, which grow at an abnormally fast rate. Because these cells generally have no other function than to divide and grow, they begin taking up the resources and space for normal cell growth. From the point of the Central Dogma, cancer cells are replicating quickly as well as transcribing quickly. However, both phases are completely uncontrolled. A tumor is formed when the cells clump together in a mass.

Thus, in each of the preceding examples, the Central Dogma was used to explain: (1) cellular production of proteins; (2) cellular division for growth; and (3) unregulated cellular mechanisms.

DRUG DISCOVERY

Since there is a considerable amount of law concerning bioinformatics tools for drug discovery, it is useful to understand where they fit in.

In the days of the shaman, particular roots or plants were given to treat ailments. Presumably this was done as trial and error. Person 1 has a fever and is given and ingests a red orchid. Person 1 immediately dies. Person 2 has the same fever and ingests a blue orchid. Person 2 eventually gets better, but no faster than others who were given nothing. Person 3 has the same fever and ingests an orange orchid. Person 3 is cured.

This is precisely how drug discovery is performed today. Most of our medicines originate as natural products in the environment. They are refined for treatment through medicinal chemistry (that is, made into a pill) and then tested. Patients either: (1) die; (2) experience no change; or (3) get well.

An example of bioinformatics in action is the development of a treatment for a fungal infection. The goal is to kill the fungus but not the patient. You could give the patient cyanide and it would probably stop the infection, but it would also have serious and fatal consequences for the patient. It probably isn't the best idea.

After isolating the fungus, a researcher might examine its DNA. This could be done by sequencing the DNA or isolating the mRNA and sequencing that. The advantage of sequencing the mRNA is that it is known that the gene is being expressed. Once there is a list of sequences, the human genome sequence is queried. The goal is to find genes that are *not* in the human genome. Once a list is created, a test is performed by "knocking out" genes on the list and noting fungal death. The results show which gene(s) to target for therapy. A screening test, where the gene is extracted and inserted to a molecular tool, is created. For example, one could insert the gene into a longer DNA sequence, which if expressed would interact with a dye and turn the test tube blue. Those test tubes that do not turn blue indicate that something has disrupted the gene.

Thousands, if not hundreds of thousands, of molecules are tested in this manner. Those that have the desired effect are studied more closely to understand the sequence and possibly even the structure of the molecule. The molecule(s) that look most promising as a treatment are sent off to the medicinal chemistry department. It is there that chemists work to modify the molecule to increase cellular uptake and lethality to the target. Once a molecule has been modified, the "drug" can then be tested.

Several legal issues that relate back to bioinformatics are incorporated in this drug development process. First, there may be intellectual property (IP) in the tools used to test gene expression. Second, there may be IP in the tools used to create and analyze the results of the screening. Third, there may be IP in the tools used to separate molecules from mixtures. Fourth, there may be IP in the tools used to understand the protein structure. Finally, although not related to bioinformatics, there is IP (usually compensation of matter) related to the drug.

HUMAN GENOME PROJECT

Because the field of bioinformatics grew exponentially in terms of utility and scientists during the Human Genome Project, it is worthwhile to understand the project as a backdrop to the growth of databases and software tools.

A common myth is that the Human Genome Project originated at the National Institutes of Health (NIH) in 1990. The Human Genome Project is actually rooted in the dropping of atomic weapons during World War II.[6]

> The background of this program begins shortly after Japan's surrender, when a Joint Army-Navy Commission made extensive observations in Hiroshima and Nagasaki on the survivors of the bombings. At the conclusion of the Commission's work its chairman, Col. A.W. Oughterson, M.C., AUS, recommended to the Surgeon General of the Army that the Council be requested to undertake a long-range study of the medical and biological effects of the atomic bomb, and this recommendation was transmitted by Surgeon Gen. Norman T. Kirk to Lewis H. Weed, chairman of the Division of Medical Sciences. As a result, in June 1946 a conference group was convened by the [National Research] Council, and in November, following its recommendation, a five-man commission composed of representatives of the Council, the Army, and the Navy left for Japan for the purpose of determining the current status of Japanese work on atomic bomb casualties, evaluating the feasibility of American participation in continued research on these casualties, and indicating the lines along which such studies should proceed. This commission, known as the Atomic Bomb Casualty Commission . . . submitted a report of its findings to the Council in January 1947.[7]

Trying to understand the effects of radiation on those exposed did not stop with those directly injured but rather carried on into their offspring. Thus, abnormalities shown in offspring were thought to be indicative of radiation exposure of the parent(s). Several decades later, the Human Genome Project itself was directly addressed:

> With these capabilities, OHER [Office of Health and Environmental Research] began in 1986 to consider the feasibility of a dedicated human genome program. Leading scientists were invited to the March 1986 international conference at Santa Fe, New Mexico, to assess the desirability and feasibility of implementing such a project. With virtual unanimity, participants agreed that ordering and eventually sequencing DNA clones representing the human genome were desirable and feasible goals. With the receipt of this enthusiastic response, OHER initiated several pilot projects. Program guidance was further sought from the DOE [U.S. Department of Energy] Health Effects Research Advisory Committee [HERAC].[8]
>
> The HERAC Recommendation. The April 1987 HERAC report recommended that DOE and the nation commit to a large, multidisciplinary, scientific, and technological undertaking to map and sequence the human genome. DOE was particularly well suited to focus on resource and technology development, the report noted; HERAC further recommended a leadership role for DOE because of its demonstrated expertise in managing complex and long-term multidisciplinary projects involving both the development of new technologies and the coordination of efforts in industries, universities, and its own laboratories. Evolution of the nation's Human Genome Project further benefited from a 1988 study by the National Research Council (NRC) entitled Mapping and Sequencing the Human Genome, which recommended that the United States support this research effort and presented an outline for a multiphase plan.[9]

NIH was later brought into the program, as were numerous developed nations (e.g., the UK, France, and Japan).

> The National Institutes of Health (NIH) was a necessary participant in the large-scale effort to map and sequence the human genome because of its long history of support for biomedical research and its vast community of scientists. This was confirmed by the NRC report, which recommended a major role for NIH. In 1987, . . . NIH established the Office of Genome Research in the Director's Office. In 1989 this office became the National Center for Human Genome Research (NCHGR), directed by James D. Watson. . . . In 1997 NCHGR became the National Human Genome Research Institute (NHGRI).[10]

So where does bioinformatics fit in? The first place to look is in the goals of the Human Genome Project.

Goal 1: Map All Human Genes

During this part of the Human Genome Project, bioinformatics use was focused on the development of databases (both to store raw data necessary for the task and to present finished results). The software during this period, much like the databases, was rudimentary by today's standards. The software focused on ways to understand human pedigrees and follow genetic diseases through families, as an example.

Goal 2: Determine the Human Genome DNA Sequence

This event is the bioinformatics equivalent of the universe's Big Bang. First, databases had to be created to store the vast amount of data that DNA sequencing produces. With a sequence of 3.1 billion bases, and with a coverage of approximately five-fold needed to piece the small sequences together, databases that could hold and display 15.5 billion bases were needed. The reason five-fold coverage is needed will be discussed later.

In addition to the databases, software was needed to assemble the human DNA sequence. Additionally, enormous computing power was needed.

Goal 3: Determine the Function of All Human Genes

This final goal of the Human Genome Project will most likely never be met. For several years after the human sequence was published, there were debates on exactly how many human genes existed.[11] Estimates went up to 120,000! It was finally settled that the human genome is made up of approximately 20,000 to 25,000 genes.[12] To determine the function of all these genes would be impossible. However, the first to be determined were single-gene diseases (e.g., cystic fibrosis, phenylketonuria, and Duchenne muscular dystrophy). These diseases occur if there is a mutation in a single gene. The body of law around genes is enormous. Most familiar is the *Association for Molecular Pathology v. U.S. Patent & Trademark Office*,[13] in which the courts were asked to determine what, if any, intellectual property rights are attached to genetic tests for a disease (breast cancer). Again, as the bio-informatics community works to unravel the functions of genes, databases and analytical tools present the most pressing legal questions.

Most people involved are retired, have left the field, or are dead. When the human genetic map was being assembled, chromosome "editors" throughout the world met yearly to go over the data and draw the maps. Eventually computers were brought in to deal with the increasing amount of data, as well as to allow for easy access and continuous curation of genetic maps. The GDB Human Genome Data Base was the official repository for this information.[14]

As the information grew and the database became more integral to the Human Genome Project, scientists (and governments) outside the United States were concerned that: (1) their work was not appreciated, and the Human Genome Project was viewed solely as a U.S. project; and (2) the United States could cut off access to the database. While both of these assertions may seem ludicrous in hindsight, they were legitimate concerns of scientists.

This resulted in the creation of GDB nodes.[15] A node was a copy of GDB placed at a remote site (e.g., Japan). The technology at the time (given the funds available for the project) did not allow for the nodes to accept data and transfer it to GDB. However, each node could download the daily updates to the database. Of course, the collection of all the data was in the public domain.

THE SEQUENCE

Coinciding with the Internet bubble, there was a bioinformatics bubble. The bubble concerned who would be the first to have the human DNA sequence. What was later referred to as the "Public Project" (involving NIH, DOE, European Commission, etc.) was following a logical and stepwise progression. The first step was to break down the parts of genome into smaller pieces and sequence these pieces or walk across a piece of DNA as it is sequenced. While doing so, funding would be allocated for the development of new sequencing technology.

Then all hell broke loose! A private company, Celera, announced that it was sequencing the human genome. Celera also was doing it in a very different way. The technique it used was called shotgun sequencing.[16] In shotgun sequencing, one makes multiple copies of the genome and randomly breaks them apart. In order to get enough overlapping pieces for assembly, this usually means sequencing four- to ten-fold.

The scientific community arose like villagers in *Frankenstein* to attack this method of sequencing, as well as the fact that a private company was cutting in on an internationally coordinated project along with its government funding. The decision was made in NIH to "beat" Celera to the release of the DNA sequence. The reasoning, in part, was a fear that the U.S. Patent and Trademark Office and the courts would allow the sequence to be patented.[17] One company having a monopoly on the human DNA sequence would change the face of human disease research.

Luckily, that did not happen. The human genome sequence remains in the public domain and Celera holds its own sequence privately.[18] A joint press conference with President Bill Clinton and the heads of the public and private efforts took place on June 26, 2000.[19]

In the end, the private effort concentrated on selling the results of analyses (bioinformatics) and variations (by looking at regions in multiple populations). Because human sequence varies approximately every 10,000 bases, and with the exception of identical twins, no two people have the same sequence, the DNA sequences of the public and private efforts had variability. This variability resulted in the HapMap Project, whereby a mapping of DNA patterns is internationally funded.[20]

DATABASES

There are a number of useful databases. For the sake of completeness, here are a few of the most used:

NCBI (National Center for Biotechnology Information) http://.www.ncbi.nlm.nih.gov	DNA Sequences & Associated Data
PubMed http://.www.ncbi.nlm.nih.gov/pubmed/	Literature
OMIM (Online Mendelian Inheritance in Man) http://.www.ncbi.nlm.nih.gov/omim/	Genetic Medicine
UniProt (Universal Protein Resource) http://.www.uniprot.org	Universal Protein Knowledgebase
GEO (Gene Expression Omnibus) http://www.ncbi.nlm.nih.gov/geo/	Gene Expression
PDB (Protein Databank) http://www.rcsb.org/pdb/home/home.do	Protein Structures
International HapMap Project http://hapmap.ncbi.nlm.nih.gov/	Genetic Variation

Each of these databases provides its information for *free*, unlike many of the databases used by the legal profession. The reason these databases are free, however, is that federal and international funding provided hundreds of millions (if not billions) of dollars to produce them. Some databases, in fact, allow people to download the entire set of data without so much as a log-in.

The National Center for Biotechnology Information, listed above, hosts numerous databases and software tools. It takes a very relaxed position on intellectual property:

> Copyright Status:
>
> Information that is created by or for the US government on this site is within the public domain. Public domain information on the National Library of Medicine (NLM) Web pages may be freely distributed and copied. However, it is requested that in any subsequent use of this work, NLM be given appropriate acknowledgment.[21]

This is not to say that all of the data can be repackaged for commercial gain. For example, OMIM states the following:

> The OMIM® database including the collective data contained therein is the property of the Johns Hopkins University, which holds the copyright thereto.
>
> The OMIM database is made available to the general public subject to certain restrictions. You may use the OMIM database and data obtained from this site for your personal use, for educational or scholarly use, or for research purposes only. The OMIM database may not be copied, distributed, transmitted, duplicated, reduced or altered in any way for commercial purposes, or for the purpose of redistribution, without a license from the Johns Hopkins University.[22]

When a database is being used, especially for commercial purposes, it is advised that the intellectual property position of the database owner/host be examined. In many cases, databases may be semi-public. The most notorious example occurred in 1994. In that year, The Institute for Genome Research (TIGR) placed restrictions on academic use of its data. Most notable was the requirement for academic researchers to send their manuscripts prior to publication to a TIGR-affiliated for-profit business. This allowed the staff to determine whether or not a particular region of the genome would be worth patenting.[23]

When it comes to issues of use and attribution, there are two broad categories of bioinformatics data. The first is Database-Prime. The use of this term is to distinguish it from other databases, which may be an amalgamation of data from other sources. One example of a Database-Prime is PAHdb (phenylalanine hydroxylase database).[24] This is a heavily curated source of primary data on genetic information relevant to phenylketonuria. The database lists no fewer than a dozen people involved with it.

As a Database-Prime, it is not surprising that PAHdb takes a slightly stronger approach to protecting its intellectual property:

> COPYRIGHT:
>
> PAHdb constitutes the intellectual property of the McGill University Health Centre and its collaborating partners (DeBelle Database Group). Any unauthorized copying, reutilization, extraction, storage, or distribution of this material or a substantial part thereof without written permission from the PAHdb Curators would lead to copyright and database right infringement with possible ensuing litigation. Copyright and database right of PAHdb. All rights reserved.[25]

In contrast, resources such as those amalgamated at NCBI have a much looser intellectual property position.

Finally, there is a sort of underground intellectual (or nonintellectual, depending on one's view) property system developed by scientists regarding data and lab samples relevant to bioinformatics and its supported disciplines. The result of this system is either the inclusion of authors on a publication who contributed little, if anything, to the research, or the exclusion of the bioinformatics team from the publication. Since the mantra of "publish or perish" is clearly shouted in the halls of academe, scientists look for any way to list a publication on their curriculum vitae. This results not only in publication inflation on the part of the scientist, but also in author lists with over 100 authors on scientific publications.

A common way in which scientists make this happen is by sharing a sample. A scientist might have a sample of DNA on which another scientist wishes to perform a test. Scientist A mails the samples and, voilà, Scientist A is now on a paper. Although the journals are trying to discourage this blatant disregard for scholarly effort by requiring all scientists to sign documentation that they played a part in the research and/or drafting of the publication, the situation is still far from optimal.

Conversely, scientists also see the need to decrease the number of authors of the papers in order to make them appear more important to the research. However, in this situation it is the bioinformatics team that regularly gets removed from authorship. Scientists are known to justify this based on the fact that "there was no intellectual novelty; the bioinformatician just ran programs on my data." *How would this contribution be distinguishable from mailing some test tubes?*

The goal here is not to make a case alleging that the peer-review system is broken, but rather to point out that the cited work of bioinformaticians on papers is just a portion of the work that groups have performed. While there are clear methods to deal with plagiarism, it is a muddy situation to deal with lack of attribution; and in many of these cases taking such an action could mean professional suicide.[26]

SOFTWARE

After data, software is the other broad area of bioinformatics. While it is good to collect data, it is the software that turns the data into knowledge. Legal issues concerning software are addressed elsewhere in this book. Therefore, this section will present an overview of the field.

The first major software program in bioinformatics that performed sequence alignments was FASTA (FAST-All). Initially it was used primarily for DNA sequencing, but the software supports protein sequencing as well.[27] The software was so fundamental to the field that sequence databases would export data in FASTA format. Later a tool with improved mathematic algorithms, BLAST (Basic Local Alignment Search Tool), was released.[28]

In each of these cases, the software was released into the public domain. This allowed other groups to incorporate the tool and its algorithm in packaged "Bioinformatics Portals," such as the one provided at NCBI. While it is a commonly known fact that one cannot patent mathematics, that would not preclude the inventors from attempting to patent (or copyright) the software.

The importance of alignment tools cannot be overstressed. As previously discussed, DNA sequencing involves the use of alignment tools to piece together the short pieces into a longer DNA molecule. However, there are many more uses for alignment tools.

While bioinformatics is most commonly associated with the field of genetics (genomics), it also plays a key role in the fields of population and evolutionary science. If one assumes that an organism accumulates mutations in a gene (those that do harm the organism), then organisms whose DNA sequence is most similar to another are more closely linked in an evolution perspective. Much of evolutionary biology concerns itself with building "trees."

Such trees might show the development of entire organisms or the evolution of a single gene. For example, if one wishes to look at the evolutionary history of a gene, a protein is selected that has strong variability across the examined species. These sequences are aligned, additional mathematics may be used, and ultimately a tree can be constructed.

As discussed, proteins are made up of a string of amino acids, which fold into an active structure. An entire branch of bioinformatics is devoted to predicting the process of protein folding. This is a very difficult process. Trying to determine the "best" tool is difficult due to the number of variables that can be tested; however, the best predictors have an "accuracy" of approximately 67 percent.[29]

Next to trying to assemble an entire genome, protein folding requires the second greatest amount of computational resources. Scientists have been ingenious in getting the community involved by creating "Folding@Home," a software program that runs when a computer is idle, and works a region of protein.[30]

The license agreement for Folding@Home seems lacking:

> License Agreement:
>
> Please carefully read the following terms and conditions before using this software. Use of this software indicates acceptance of this license agreement and disclaimer of all warranties.
>
> Disclaimer of Warranty:
>
> IN NO EVENT SHALL STANFORD UNIVERSITY BE LIABLE TO ANY PARTY FOR DIRECT, INDIRECT, SPECIAL, INCIDENTAL, OR CONSEQUENTIAL DAMAGES, INCLUDING LOST PROFITS, ARISING OUT OF THE USE OF THIS SOFTWARE AND ITS DOCUMENTATION, EVEN IF STANFORD UNIVERSITY HAS BEEN ADVISED OF THE POSSIBILITY OF SUCH DAMAGE.
>
> STANFORD UNIVERSITY SPECIFICALLY DISCLAIMS ANY WARRANTIES, INCLUDING, BUT NOT LIMITED TO, THE IMPLIED WARRANTIES OF MERCHANTABILITY AND FITNESS FOR A PARTICULAR PURPOSE. THE SOFTWARE AND ACCOMPANYING DOCUMENTATION PROVIDED HEREUNDER IS PROVIDED "AS IS". Folding@Home HAS NO OBLIGATION TO PROVIDE MAINTENANCE, SUPPORT, UPDATES, ENHANCEMENTS, OR MODIFICATIONS.
>
> Restrictions:
>
> You may use this software on a computer system only if you own the system or have the written permission of the owner. You may not alter the software or associated data files, or de-compile or reverse engineer the software. You may only use unmodified versions of Folding@Home obtained through authorized distributors to connect to the Folding@Home servers. Use of other software to connect to the Folding@Home servers is strictly prohibited. This prohibition includes 3rd party installers, which download directly from Stanford web sites, unless written permission is granted from Stanford University. Distribution of this software is prohibited. It may only be obtained by downloading from Stanford's web site (*http://folding.stanford.edu* and pages linked therein) or the web site of one of our commercial partners (Sony, NVIDIA, and ATI).[31]

While the license is nondescript, it does not mention ownership of the results! While tens of thousands (if not hundreds of thousands) of computers work to fold a single protein, each user is contributing to that effort. The legal question is whether a user has any rights in the results of his or her work on this project. This will be addressed later in the book.

ANNOTATION SOFTWARE

The final class of software discussed here is annotation software. It is through annotation that individuals, groups, and even computers can add information to data already stored in an existing database.

The simplest form of annotation occurs when software examines a piece of data and places additional information in the database or in a linked database. The human sequence displayed at NCBI has been annotated by computer to predict the regions containing genes.[32] NCBI provides an overview:

> As genome sequence data become available for an organism, NCBI staff work to provide the data as a reference sequence (RefSeq) for display in the Map Viewer. We have developed several protocols to reach this goal and rely heavily on collaboration with genome-specific research groups whenever possible. NCBI provides various levels of computation, analysis, and curation as needed per organism. For instance, most genomes are assembled by an external group and annotated via the NCBI annotation pipeline (e.g. rat, bee, fly). For the human and mouse genomes, NCBI computes the assembly in collaboration with the international sequencing consortium; NCBI and other external groups independently provide annotation on the assembled genome. And, for other genomes, such as Drosophila melanogaster, the NCBI RefSeqs represent the assembly and annotation as provided by the fly sequencing consortium.
>
> NCBI is providing reference sequence (RefSeq) records that represent assemblies of genomic sequence data and the corresponding RNA and protein sequences. For externally provided assemblies, there is no guarantee that the genomic RefSeq will correspond exactly to the submitted assembly as contaminants will be screened out in the RefSeq version (see the **Assembly** page for more information). The NCBI annotation pipeline annotates the genomic RefSeq data with features such as genes, RNAs, proteins, variation (SNPs), STS markers, and FISH mapped clones. All sequences (genomic, RNAs, proteins) are available for customized **BLAST** searches. BLAST results, as well as the sequence features, are readily displayed on NCBI's **Map Viewer**.[33]

NCBI takes a dual model of annotation. It not only uses its software but allows other groups to add information as well. While NCBI is not set up for single-person annotation, it does provide for annotations submitted by organizations that it recognizes as being of high scientific quality.

However, NCBI is experimenting with third-party annotation in its TPA:experimental Sequence Database. Here NCBI states:

> Published wet bench experiments are the most critical and valuable data to validate, add, or correct annotation on records in GenBank, particularly for those annotations on genome records that are largely computational results. The TPA:experimental database was established specifically to capture these experimental results in a familiar format. In addition, a publicly accessible TPA record will be linked to a publication that documents that the data are supported by biological experimentation.[34]

There is no specific text asserting that submitters of data give up all rights, or that data is automatically placed in the public domain. However, since that is the general assertion for most NCBI databases, one would be hard pressed to argue otherwise.

CONCLUSION

Understanding bioinformatics is difficult, primarily because of the vastness of what it may or may not include. As Supreme Court Justice Potter Stewart said of obscenity, "I know it when I see it." Not to be flippant, but that is the state of bioinformatics; some areas clearly (for most) are bioinformatics, and others are mathematics, statistics, computer science, etc. And some are a combination of areas.

Law is reactive in its nature. Issues are addressed by the legislature when there is an outpouring of need for action. The courts act when there is standing to bring forth a new issue of law. It is unlikely that there will be any laws passed concerning bioinformatics. Although bioinformatics concerns may be addressed either implicitly or explicitly in laws concerning issues such as genetic privacy, other than the Tax Court, there is no record of adjudication on bioinformatics.

Therefore, a legal practitioner must rely on applying existing laws to a relatively new discipline. Issues such as intellectual property, tort liability, design, and so on have a rich body of law. It is hoped that this book will provide the practitioner with at least a starting point in examining legal principles relevant to bioinformatics.

NOTES

1. N.M. Luscombe et al., *What Is Bioinformatics? A Proposed Definition and Overview of the Field*, 40 METHOD INFO. MED. 346 (2001).
2. Paulien Hogeweg, *The Roots of Bioinformatics in Theoretical Biology*, 7 PLOS COMPUTATIONAL BIOLOGY, no. 3, 2001 at e100201.
3. T.C. Summ. Op. 2007-96 (June 14, 2007).
4. A. JAMES CUTICCHIA, GENETICS: A HANDBOOK FOR LAWYERS (2010).
5. *Id.*
6. NATIONAL RESEARCH COUNCIL, THE CHILDREN OF THE ATOMIC BOMB SURVIVORS: A GENETIC STUDY (James V. Neel et al., eds., 1991), *available at* http://www.nap.edu/openbook.php?record_id=1800&page=9.
7. *Id.*
8. U.S. DEPARTMENT OF ENERGY 1991–1993 HUMAN GENOME PROGRAM REPORT (June 1992), *available at* http://www.ornl.gov/sci/techresources/Human_Genome/project/91history.shtml.
9. *Id.*
10. *Id.*
11. Bijal Trivedi, *How Many Human Genes?*, GENOME NEWS NETWORK (May 26, 2000), *available at* http://www.genomenewsnetwork.org/articles/05_00/how_many_genes.shtml.
12. *How Many Genes Are in the Human Genome?*, HUMAN GENOME PROJECT (Sept. 19, 2008), http://www.ornl.gov/sci/techresources/Human_Genome/faq/genenumber.shtml.
13. 653 F.3d 1329 (Fed. Cir. 2011).
14. P.L. Pearson, *The Genome Database (GDB)—A Human Gene Mapping Repository*, 19 NUCLEIC ACID RES. Suppl. 2237 (1991).
15. A. James Cuticchia et al., *The GDB Human Genome Database Anno 1993*, 21 NUCLEIC ACIDS RES. 3003 (1993).
16. P.C. Ng & E.F. Kirkness, *Whole Genome Sequencing*, 628 METHODS MOLECULAR BIOLOGY 215 (2010).
17. D.L. McKay, *Patent Law and Human Genome Research at the Crossroads: The Need for Congressional Action*, 10 SANTA CLARA COMPUTER & HIGH TECH. L.J. 465 (1994).
18. Eric Lander et al., *Initial Sequencing and Analysis of the Human Genome*, 409 NATURE 860 (2001); J.C. Venter et al., *The Sequence of the Human Genome*, 291 SCI. 1304 (2001).
19. Jamie Shreeve, *The Blueprint of Life*, U.S. NEWS & WORLD REP. (Oct. 31, 2005), http://www.usnews.com/usnews/news/articles/051031/31genome.htm.
20. International HapMap Consortium, *The International HapMap Project*, 426 NATURE 789 (2003).
21. *Copyright and Disclaimers*, NCBI, http://www.ncbi.nlm.nih.gov/About/disclaimer.html (last visited Feb. 15, 2012).
22. *Copyright Status*, OMIM, http://www.ncbi.nlm.nih.gov/Omim/restrictions.html (last visited Feb. 15, 2012).
23. KEVIN DAVIES, CRACKING THE GENOME: INSIDE THE RACE TO UNLOCK HUMAN DNA (2001).
24. http://www.pahdb.mcgill.ca (last updated Aug. 31, 2009).
25. *Terms of Use and Copyright Notice*, PAHdb, http://www.pahdb.mcgill.ca/?Topic=About&Section=Copyright&Page=0 (last visited Feb. 15, 2012).
26. D.R. Forsdyke, *Authorship and Misconduct*, 370 NATURE 91 (1994).
27. W.R. Pearson & D.J. Lipman, *Improved Tools for Biological Sequence Comparison*, 85 PROC. NAT'L ACAD. SCI. U.S.A. 2444 (1988).
28. S.F. Altschul et al., *Basic Local Alignment Search Tool*, 215 J. MOLECULAR BIOLOGY 403 (1990).

29. Andriy Kryshtafovych et al., *Evaluation of Model Quality Predictions in CASP9*, 79 Proteins: Structure, Function, & Bioinformatics, no. S10, 2011 at 91.

30. Adam Beberg & Vijay S. Pande, *Folding@Home: Lessons from Eight Years of Distributed Computing*, IEEE International Symposium on Parallel & Distributed Processing 1–8 (2009), *available at* http://doi.ieeecomputersociety.org/10.1109/IPDPS.2009.5160922.

31. *Folding@Home Distributed Computing Client*, Folding@Home, http://folding.stanford.edu (last visited Feb. 15, 2012).

32. G.D. Schuler, *Sequence Mapping by Electronic PCR*, 7 Genome Res. 541 (1997).

33. *Annotation Information*, http://www.ncbi.nlm.nih.gov/projects/genome/guide/build.shtml (last updated Oct. 15, 2012).

34. *TPA:experimental Sequence Database*, http://www.ncbi.nlm.nih.gov/genbank/TPA-Exp (last updated Aug. 27, 2012).

CHAPTER 2

Commercial Software Licensing

Jeremy T. Marr

INTRODUCTION

Licensing commercial software may be a preferable alternative to in-house development or software as a service (SaaS) for many organizations looking to obtain access to bioinformatics capabilities.

Some of the benefits of commercial software packages are that they do not require investing in developing custom software, they typically provide cost-effective options for ongoing support and enhancement, and they can provide access to needed features quicker than those features can be developed in-house (even if in-house developers make liberal use of open source code). Commercial licensing may also be preferable to acquiring the right to access bioinformatics software through an SaaS model in certain circumstances. For instance, you may prefer to license commercial software if needed features are not available under the SaaS model; if you are concerned about the security of your data under such a model (including maintaining the secrecy of the particular sequences on which your research is focused); if you have a robust internal IT infrastructure and find it more cost-effective to acquire a license than to pay recurring access fees; if the commercial vendors are more established than the SaaS vendors; or if the transmission of immense volumes of data between your computers and the SaaS host would render the SaaS model too slow and ultimately unworkable for using certain features.

SELECTING COMMERCIAL BIOINFORMATICS SOFTWARE

Once you have decided to obtain a license to commercial bioinformatics software, you must select the software that best meets your needs. Key factors to consider in making this selection are the features provided by the software, including the software's compatibility with your organization's bioinformatics platform; the capabilities of the software vendor, including the vendor's ability to provide the services that the organization will need; and the available licensing models and corresponding costs.

Features

The first step in selecting a commercial software package is to identify vendors that provide software having the analysis, database, search, visualization, reporting, project management, and other functionality you need. Most vendors provide detailed descriptions of their software on their websites that can help you initially screen for available options. Since it is often difficult to get a solid feel for software by reading such materials, which can be laden with marketing claims, you will probably want to obtain a trial copy of the packages that interest you most so that you can verify that the features you need will work efficiently and in the way you need them to work. Most vendors will be open to letting you use their software for a limited evaluation period, particularly if you have a budget allocated and a concrete plan to compare competitive packages.

Compatibility

You will want to be sure that the bioinformatics software packages you select are compatible with one another and with the rest of your computing infrastructure. In particular, you will want to be sure that the software works with your preferred public and private databases, file formats, and any productivity software you use, such as word processing and presentation packages.

Vendor Capabilities

The services available to you will vary from vendor to vendor. Most vendors will provide technical support for their software package through some combination of e-mail, their websites, and telephone access. This service is often included with the fees you pay for licensing the software package, depending on the licensing model (as discussed later in this chapter). Vendors will also provide software maintenance, which usually consists of updates (bug fixes, minor enhancements, data updates, and ongoing compatibility) and upgrades (major enhancements) to the software package, either included as part of the licensing agreement or for a separate fee. As you screen vendors, you should determine each vendor's typical release cycle for updates and upgrades, particularly if you have any concerns about the current version of the package.

Many vendors will offer training and other professional services, including assistance with installing, customizing, and implementing their software packages. Vendors typically charge fees for such services over and above the fees they charge for the software packages, and they may also restrict the ways in which any customized code can be used. You should select vendors that offer the services you think you will need and can afford.

License Options and Costs

As a preliminary matter, it is important to note that, unlike other copyrighted material such as books and music, copies of software are not sold. Instead, they are licensed under the terms of an agreement that, among other things, describe permitted and prohibited uses. Use of the software beyond the terms, or scope, of the license can give rise to an infringement claim. Vendors provide access to their software packages using a number of different licensing approaches, which is the subject of the next section of this chapter.

LICENSING MODELS AND SCOPE

Commercial software development is an expensive undertaking, involving the work of teams of skilled developers over several months or years. Vendors seek to recoup their investment in creating software packages by licensing them for a fee. Since a vendor incurs very little additional cost in providing additional users with access to the software, the fees charged do not typically bear any close relationship to the vendor's cost in providing the customer with a copy of the software, which is the cost-based pricing model commonly used for goods with high per-unit manufacturing costs. Instead, vendors typically use a value-based model, where the fees charged by a vendor are designed to be proportional to the value the customer receives from the software. The value proposition offered by software vendors is typically based on models that demonstrate that the customer will enjoy substantial savings by using the software, as compared to other approaches the customer might take to perform the functions that the software performs.

Scope of Use

Software vendors offer several types of licenses, each having a different scope of use.

Enterprise Licenses

The traditional, unlimited license scope is often termed an "enterprise" license. Enterprise licenses allow for unlimited use of the software throughout an organization, regardless of the number of users, number of computers, or locations involved. As one would expect, enterprise licenses are usually the most costly type of license, with the price typically determined based on the size of the organization. However, enterprise licenses may be a good solution for university systems that are made up of several independent colleges.

Site Licenses

One step down in scope from an enterprise license is what is often termed a "site" license. Site licenses are similar in most ways to enterprise licenses, except that they are limited to use at a specified location, such as a single campus or office address. The geographical limitation imposes a practical limit on the number of users.

Per-Computer Licenses

Vendors also offer licenses on a per-computer basis. In this model, a customer purchases a license for each computer on which it wishes to install and use the software. These used to be termed "CPU licenses," but these days even notebook computers come with multiple CPUs. Some software packages include "license management" software designed to ensure that the software is installed and used only on the computers for which it is licensed. Problems with this model can arise if a computer needs to be replaced; most vendors provide a means for obtaining an updated "license key" to enter into the license management software, which allows for transfer of a license from one computer to another in such situations.

Named-User Licenses

Two user-based licensing models are commonly used, and the same vendor may offer both. The first is often termed a "named-user" license. Software licensed in this manner can be accessed from any number of computers, but the license management software ensures that the software is used only by a specified set of users identified by name or other unique credentials. Vendors typically provide a means for updating the list of authorized users from time to time to account for staff turnover, and they also provide a means for adding additional users following payment of a per-user upgrade fee.

Concurrent-User or Per-Seat Licenses

The second user-based licensing model addresses not only the staff turnover issue, but also the fact that not every named user will use the software the same amount of time or even concurrently. The "concurrent-user" or "per-seat" model allows for licensing of a specified number of concurrent users without regard to their identity or location. The license management software will ensure that no more than the licensed number of users are using the software at any given time. Some software will allow for limited use in excess of the licensed number of users, accompanied by gentle (or not-so-gentle) reminders that the purchase of additional "seats" may be in order. Because of the added flexibility of per-seat licensing, and the fact that each seat may be used by more than one user (albeit not at the same time), a license for, say, 10 seats is typically more expensive than a comparable license for 10 named users.

Metered Licenses

Metered licenses are a less-traditional type of software license. Vendors offering these types of licenses seek to tie the cost of the license more directly to the value the customer receives from using the software, similar to how the amount a homeowner pays for electricity is based on the amount of electricity consumed. These licenses can be priced similarly to prepaid cellular phone service, with the customer paying in advance for a specified level of use. Once that use is consumed, the consumer would "reload" the software by purchasing additional usage units. The usage units vary from software package to software package, based on the main functionality performed by the software. For bioinformatics software, such a model may be based on the number of DNA sequences worked with, the number of sequence analyses performed, the number of similarity searches performed, or the number and type of databases accessed.

Perpetual Licenses

The traditional mode of licensing software entails the customer purchasing what is termed a "perpetual" license. This means that the customer, in exchange for a single, usually upfront, license fee, obtains the right to use the software indefinitely, subject to the customer's compliance with the terms of the license. No additional license fee is ever due, although vendors often combine perpetual licenses with an annual maintenance fee that covers the provision of technical support services and access to updates and new versions of the software (or "maintenance").

Term/Subscription

As an alternative to perpetual licenses, some vendors offer software packages under licenses that last for specified periods of time. As would be expected, these licenses are less expensive than perpetual licenses. However, the customer's right to use the software expires at the end of the license term unless the customer renews the license (and pays a renewal fee) by a specified date prior to expiration. Some vendors combine a one-year term license with one year of maintenance and support services, referring to such a bundle as a one-year "subscription."

Editions

Some software packages are offered in multiple "editions," each of which contains a specific set of features or authorized scope of use. Generally, the more fully featured, less-restrictive editions provide more robust functionality but at a greater price. Lower-end editions cost less but either contain fewer features or are subject to tighter use restrictions. Vendors hope that the lower-cost editions will attract larger numbers of users, many of which may elect to upgrade to the more costly editions in order to gain access to premium features or to avoid the limitations of the less-expensive edition.

In determining which edition of a software package to license, you should make a realistic assessment of your organization's likely needs, not just initially but over the projected period of use for such software. If it seems likely that the organization will need access to the functionality contained in a higher-cost edition, it may make sense to purchase that edition initially rather than a lower-priced edition. This is because the cost of a lower-priced edition combined with the cost of an upgrade to the higher-priced edition will typically be more expensive than the cost of the higher-priced edition. Alternatively, if the organization needs multiple copies of a package, it can purchase multiple editions and, if permitted under the licensing model, allocate them among users accordingly. Finally, an organization can try to negotiate pricing and upgrade paths with the vendor up front so that if the organization purchases the wrong edition, it will not be more expensive for it to obtain the proper edition than it would have had the organization purchased it initially.

POINTS OF NEGOTIATION

Once you have selected the software package you are interested in and have vetted the capabilities of its vendor and the different licensing terms offered, the next step will be to enter into some type of written agreement with the vendor that provides you with the necessary rights you will need to use the software package. This gives you the opportunity to negotiate various aspects of the transaction.

Most commercial software is provided under a standard end-user license agreement (EULA), the terms of which the user is prompted to accept as a precondition of using the software. The user is given no opportunity to propose alternative terms or to negotiate mutually agreed terms with the vendor. Nevertheless, the terms of EULAs, though generally vendor-favorable, are in theory kept in check by the terms offered by competitors; that is, a vendor who offers user-unfriendly terms as compared to its competition risks losing customers to the competition as a result, all else being equal. The soundness of this theory is based in part on an assumption that decision makers review and understand the terms of a EULA before committing to purchase the software—an assumption that often is not accurate.

Even where vendors use EULAs, it is often possible to negotiate different terms for large purchases. Typically, this will entail entering into a license agreement that overrides the terms of the EULA. Vendors typically reserve this approach for larger purchases because of the time and cost involved in negotiating separate agreements with each customer. The use of take-it-or-leave-it EULAs is not as common with more specialized software targeted at a smaller market. The sales model for such software often consists of a smaller number of higher value transactions, which enables the vendor's cost of negotiating license terms to be absorbed as part of its cost of sales.

Pricing

Pricing terms are not usually included in EULAs. Instead, they are typically listed on a proposal from the vendor that the customer can accept by, for example, issuing a corresponding purchase order.

The price of software licenses varies widely from software package to software package based on a number of factors, including the scope and term of the license, the features provided, bundled services, and the vendor's need to achieve a return on the investment it made in developing the software. As discussed above, unlike many products, the cost of materials is not a significant factor in pricing software. That said, most established vendors generally will not be open to discounting their software from its list price. There are, however, exceptions. For example, many vendors will offer discounts for volume sales or for nonprofit or educational customers. Vendors sometimes discount their software toward the end of a fiscal quarter as an incentive for prospective customers to purchase the software during that quarter so that the vendor can achieve its financial goals for that quarter; accordingly, it can pay to learn when end-of-quarter periods fall for a given vendor and focus purchasing decisions on those periods.

Vendors in an emerging area such as bioinformatics that are seeking to establish a referenceable customer base may be willing to provide their software at a discount in exchange for the customer's assistance in marketing efforts, even if that assistance is limited to agreeing to be listed on the vendor's website as a representative customer.

Once a customer has installed the software and is relying on it, it can become difficult and expensive to change to other packages. Vendors know this, and can sometimes increase license and maintenance fees substantially following the initial term of the license. To protect against this, you should negotiate pricing for renewal periods when you initially license the software. If you cannot agree on specific pricing, you should be able to agree that any increases will not be in excess of a certain annual percentage rate, such as the consumer price index. The vendor should have to notify you of any increase well in advance of the date you need to commit to renew the license so that, if you find the increase excessive, you will have time to seek alternatives.

Warranties/Indemnification

Software vendors typically warrant only that the software will materially conform to its documentation. Vendors usually will not offer and will disclaim any warranty of noninfringement because it is impossible to know at any given time of all potential claims any third party might make (some of which might be legitimate) that the software or its use infringes or misappropriates intellectual property rights. For example, a third party may have an unpublished patent application pending that covers certain methods that the software may employ. Instead, most vendors will agree to indemnify any customer from any third-party claims that the use of the software by the customer infringes or misappropriates the third party's intellectual property rights. A vendor will seek to resolve any such claim by its choice of the following three approaches: modifying the software so that it is no longer infringing; obtaining a license from the third party so that the customer's use is no longer infringing; or refunding to the customer a prorated amount of the applicable license fee, typically depreciated over a three-year period so that no portion of a perpetual license fee would be refunded as a result of any claim that arises three or more years after the license date. Similarly, if there is a warranty issue, i.e., the software does not work as described, the vendor will either provide an update that brings the software into conformance with its documentation, or will provide a similarly prorated refund. The logic behind prorating the refunds is that the customer was presumably able to use and obtain value from the software before the infringement or warranty claim arose, and most software becomes largely obsolete for one reason or another after three years, particularly in a fast-advancing area such as bioinformatics.

Transferability of Licenses

Software licenses are almost always nonexclusive licenses, meaning that the vendor can license any number of customers to use the software. The holder of a nonexclusive license cannot transfer the license to a third party without the vendor's express permission.[1] Most software license agreements do not provide such express authorization, and often for good reason. For instance, the vendor presumably calculated the cost of a perpetual, enterprise license based on the size of the enterprise at the time the license was purchased; if the customer transfers the license to a much larger organization, that organization would have received added benefit from the license without the vendor receiving commensurate compensation. Further, allowing for unlimited transfer of licenses can put a vendor's customers into direct competition with the vendor, compromising the vendor's ability to maintain its pricing structure. This is because an organization could purchase a license not only from the vendor but from any other organization willing to transfer its license to the software.

With this in mind, an organization should consider negotiating the right to transfer its licenses under certain specified scenarios. For example, if the software is being used to support a specific drug discovery program, it probably makes sense to negotiate the right to transfer the software license along with such discovery program.

Access to Source Code

Most vendors provide software to customers in executable form only. This is because the scope of most licenses is limited to using the software and making limited copies, and it does not include the right to modify the software or to make what is called "derivative works" under copyright law. It also serves to protect any trade secrets embedded in the source code, reducing the risk of them falling into the hands of competitors. Indeed, many customers acquire the software from the vendor in the first place because they cannot or do not wish to develop software internally. They also recognize that modifying the software will render the vendor largely incapable of providing support or further enhancements since the vendor will not have incorporated the modifications into its product or even know about the modifications.

For some organizations, however, the software may represent such a key asset that they cannot risk being fully dependent on the vendor to provide support and maintenance. Vendors can have legitimate reasons, such as the need to invest in more lucrative products, to deprioritize older products. Vendors can also go bankrupt. If a vendor is unable to or chooses not to fulfill its support and maintenance obligations, you might want the right to obtain the source code so that you, or someone you hire, can keep the software functioning.

The mechanism typically used for this purpose is a source code escrow agreement. Under such an agreement, the vendor periodically provides a third party, or escrow agent, with the latest source code for the software. The escrow agreement contains provisions requiring the escrow agent to release the source code to the customer following the occurrence of any one of a set of "release events," such as the vendor's bankruptcy or repeated failures to provide software support and maintenance services.

Improvements

Customers who purchase maintenance are typically entitled to receive updates and upgrades at no additional cost. Vendors typically reserve the right to charge extra for access to major new features, and they may charge an upgrade fee to existing customers interested in obtaining such new features.

You may want to consider negotiating the right to obtain access to any new features or database updates that the vendor releases within some fixed period of time after obtaining the license, without having to pay any upgrade fee, or, if the vendor has already told you about new features on the horizon that you particularly value, the right to get those features as part of the initial fee.

Term and Termination

With the exception of perpetual licenses, vendors typically grant you the right to use the software package only for a particular period of time. Even with perpetual licenses, vendors may provide maintenance and support services only for a limited period. If you want to extend your use of the software or services beyond this period, you will need to renew the agreement. As discussed above, to avoid surprises, you should understand the terms that will apply to renewal periods, particularly the renewal fees. It is typical for vendors to charge annual maintenance and support fees equal to 15 percent to 20 percent of the initial license fee.

To reduce transaction costs in connection with renewals, some vendors may include provisions that automatically renew the agreement for additional periods unless the customer gives notice of nonrenewal by a specific date. To avoid surprises and getting locked into the agreement for additional periods you may not need, consider negotiating a different process for renewal. For example, it would not be unusual to ask the vendor to agree to provide an invoice for the renewal period a month or two prior to the expiration date and for the customer to be able to avoid renewal by rejecting the invoice.

SUMMARY

This chapter has described the many factors that you should consider when making the decision to license a commercial bioinformatics software package. Among other things, you will need to assess the features of the package, the capabilities of the vendor, the licensing models available and their associated costs, and the other business terms proposed by the vendor to ensure that they meet your needs. With proper diligence and care, commercial software can be a valuable addition to many organizations' bioinformatics platforms.

NOTE

1. Cincom Sys. v. Novelis Corp., 581 F.3d 431, 437 n.4 (6th Cir. 2009).

CHAPTER 3

Open Source Software and Bioinformatics

Michele K. Herman, Jane Eckels, Joanne Montague, and Jordan J. Bowler

INTRODUCTION

Software in general has increasingly enabled growth and innovation in numerous scientific endeavors. The use of software in the biological and medical sciences fields, or what is referred to as "bioinformatics," is no exception. One recent observer stated:

> Not so long ago, the mention of any word with the two syllables "-ō-mics" tacked on the end was usually followed immediately with some response akin to, "Huh?" Today, we've gotten to the point where almost no biological phenomenon can escape "omics-ization," and within the next 25 years, omics will be the biggest, if not the only, game in town. Why? Because we are about to undergo what experts call a phase shift, where a technology drives a fundamental change not just in what is known, but, more importantly, in how we think of ourselves. Put another way: omics is destined to change our patterns of living in ways that only technological revolutions can deliver.[1]

Software may be developed by a single developer, multiple developers employed by one or more entities, or in an open collaborative community. Regardless of how the software is developed, it may be licensed for use by others pursuant to terms and conditions set by the software owner (the entity(ies) and/or developers who own the copyright in the software code). In recent years, there has been a trend to release bioinformatics software under license terms that allow for royalty-free access, modification, and distribution. Such licenses are referred to as open source software (OSS) licenses.

This chapter provides an overview of OSS licensing practices, the use of OSS licenses in bioinformatics, the legal risks associated with using OSS-licensed software in bioinformatics applications, and risk mitigation strategies that may be employed by scientists, researchers, academic and scientific research institutions, governmental and nongovernmental organizations, and commercial enterprises.

OSS LICENSES

Defining OSS

OSS is just like any other software except that it is licensed under a certain type of license—an OSS license. OSS licenses may be defined differently by different entities, but they share common characteristics that differentiate OSS licenses from commercial software licenses (sometimes also called proprietary licenses) in key ways. One such entity, the Open Source Initiative (OSI), maintains an Open Source Definition (OSD) that includes ten requirements that a license, in OSI's view, must satisfy to qualify as an OSS license.[2] For purposes of this chapter, however, we will refer to any license that meets the following three criteria as an OSS license:

1. The license or permission to use is self-executing;
2. There is no royalty or other monetary fee for copying, modifying, or distributing the software; and
3. The source code for the software (the "human readable form") is available to all recipients of the software.[3]

A license or permission is self-executing when the recipient is not required to sign any document or physically acknowledge acceptance or agreement. A click-through agreement would therefore not be self-executing because the end user must acknowledge his or her agreement or acceptance of the terms. Most commercially licensed software requires recipients to either negotiate and sign a license agreement or, at a minimum, accept a click-through agreement. An OSS license is more like a shrink-wrap license where one agrees to the terms of the license by opening the packaging. OSS licenses are "agreed to" simply by downloading and using the software.[4] Unlike shrink-wrap or other forms of software licenses, OSS licenses are often embedded in the software itself so that the recipient of the OSS may not even know about the license.

Types of OSS Licenses; the Spectrum of Permissiveness

There are literally hundreds of different OSS licenses, and new licenses are being developed every day.[5] While all OSS licenses are royalty-free, OSS licenses still contain significant terms and conditions to which the recipient is subject, and the variety of license terms may be bewildering. In sorting through the great diversity of OSS licenses, the various OSS licenses can be understood to be on a spectrum from the most permissive to the most restrictive. Generally, the most permissive licenses permit a recipient to redistribute the OSS under the terms of the recipient's (now an OSS licensor's) choice, i.e., its own commercial terms including monetary fees and withholding of source code, as long as the OSS licensor reproduces the copyright notice, disclaimers, and license with the OSS. In contrast, the most restrictive licenses, often referred to as "copyleft" or "reciprocal" licenses, require all changes and/or code combined with the conveyed OSS to be made available under the same OSS license and the source code to be released to all recipients of the combined code.[6] Compliance with this requirement means that anyone obtaining the combined code is entitled to obtain a royalty-free copy of the source code, making it freely available for copying, distribution, and modification. In these circumstances, there is little incentive for prospective licensees to accept a commercial license that safeguards such source code and restricts its use or distribution.[7] The Berkeley Software Distribution (BSD),[8] Artistic,[9] and MIT[10] licenses are examples of more permissive licenses. The Library General Public License (LGPL)[11] and the General Public Licenses (GPL)[12] are copyleft licenses that are examples of licenses on the restrictive end of the spectrum.[13]

INCREASING IMPORTANCE OF BIOINFORMATICS OPEN SOURCE SOFTWARE

The use and importance of OSS in the bioinformatics community has expanded significantly in the past decade.[14] Fueling this growth are the numerous benefits realized by employing open source approaches to creating and distributing software for the biological sciences over traditional proprietary models. These benefits include the rapid development of reusable software components, the broad dissemination and adoption of new tools and methods across the international scientific community, and the ability to fix bugs and generally improve the software to keep it current with emerging scientific developments.[15]

The OSS bioinformatics projects available today largely lack centralized organization or direction, despite sharing an underlying goal of promoting open source tools and methods.[16] As a result, the numerous OSS projects that have arisen have been likened to a "market bazaar" in which a variety of distinct products have been developed in response to specific needs and challenges.[17] This diversity of offerings is apparent in the wide range of OSS projects available from major bioinformatics organizations. Bioinformatics.org hosts nearly 500 active projects addressing bioinformatics issues, many of which include OSS.[18] The University of North Carolina Center for Bioinformatics, a resource for bioinformatics information, identifies approximately 50 open source toolkits provided by various developers.[19] A multitude of other institutions and umbrella organizations, such as the Open Bioinformatics Foundation,[20] develop and provide additional bioinformatics OSS projects. Included in this broad array of bioinformatics projects are, for example, educational programs, DNA analysis tools, modeling software, data integration and sharing toolkits, database management, and many others.

While the majority of developers are nonprofit entities such as universities, research institutions, and other independent collaborators, there has also been a growing commercial interest in the bioinformatics OSS environment. In 2010, Microsoft Corporation launched the Microsoft Biology Foundation (now .NET Bio), an open source, language-neutral toolkit built as an extension to the Microsoft .NET Framework.[21] Although Microsoft has since transferred ownership to the nonprofit Outercurve Foundation, Microsoft continues to support and contribute code to the project.[22] Other companies have taken a different track, opting to release both open source and proprietary versions of their bioinformatics algorithms and software. This strategy, known as "dual licensing" or "versioning," has gained popularity in the broader OSS community, in part because it allows developers to capture both users interested in the low cost or features of OSS, as well as users preferring proprietary options that enable them to improve or add code to software without being subject to open source license obligations to release such improvements or additions.[23] Some bioinformatics developers have followed suit. For example, Affymetrix, a producer of DNA microarray chips, provides both open source and proprietary versions of its software for the use of its chips.[24] Another approach, pursued by developers such as Bioclipse[25] and OpenClinica,[26] is to offer commercial support services for the bioinformatics OSS that the developer releases.

The range of licenses used in bioinformatics OSS is as wide and varied as the range of OSS projects; even so, some generalities can be noted. Although many developers in this space use customized individual licenses, much more frequently they tend to select from among the same OSS licenses that are popular in other fields. These licenses include GNU GPL, the GNU Lesser or LGPL, the Artistic License,[27] BSD type licenses,[28] and the Apache

License.[29] Some of the key features and implications of these OSS licenses are discussed further in the next section.

LEGAL RISKS ASSOCIATED WITH OSS

In the not-so-recent past, commercial enterprises, in particular, were hesitant to use OSS, and it was not uncommon for enterprises to reject out of hand the use of OSS, not just for development purposes, but in any environment. However, as use of OSS has increased, enterprises are necessarily becoming more educated regarding the risks and benefits of using OSS.

Although OSS is free in terms of royalties and fees, recipients of the OSS may have to comply with many terms and conditions. Because the licenses are embedded in the software itself, it is often difficult for an entity to learn about the license and the need to comply with its terms. Moreover, since there are so many different OSS licenses, tracking all the compliance requirements, particularly as the use of OSS increases, may be particularly challenging. An entity or developer that fails to comply with the terms of an OSS license may be subject to claims of copyright infringement and/or breach of agreement.

Since OSS is distributed freely it can exchange hands many times over, with numerous additions and changes to the code along the way. A recipient may obtain OSS that has been developed with the contributions of many individuals who may not have contributed their own code or code that they had the rights to contribute. If contributed code was unauthorized or infringes the rights of a third party, any user of that code could also be subject to claims of infringement.

While any software, irrespective of licensing, could be subject to third-party patent claims, some falsely assume that OSS is not subject to any third-party patent licenses because the OSS license itself is royalty-free. It may be that those distributing the OSS cannot sue downstream users for patent infringement but there is no reason to assume that third parties that have not written the software, used it, or distributed it would have any obligation to license their patents for free. The risk of using software that infringes third-party patents may be exacerbated with OSS since OSS, unlike proprietary licenses, will generally not include representations, warranties, or indemnification protection with regard to patent infringement claims.

OSS License Compliance

Given the often-buried nature of OSS licenses and the vast array of terms and conditions associated with them, many may wonder about the consequences of failing to comply with OSS license terms and conditions.

Recipients and OSS licensors that fail to comply with the terms and conditions of an OSS license may be copyright infringers or may be in breach of the OSS license or both. Noncompliance may result in (1) monetary damages, such as actual damages or statutory damages, (2) being enjoined from using the OSS, and (3) being required to specifically perform under the OSS license.

With regard to the financial risk of noncompliance, because OSS is released royalty-free, most view the likelihood of the copyright owner obtaining significant actual damages in the form of lost profits or the infringer's sales as being negligible. The bigger risk is the potential for statutory damages, which, if available, may be up to $30,000 for each registered work infringed.[30] Statutory damages may be increased to up to $150,000 for each registered work infringed when the infringement is found to be willful.[31] If each OSS file in a software program qualifies as separate "work" for purposes of awarding statutory damages, and the infringing program includes hundreds of such files, then the statutory damages may become substantial.[32]

An injunction to stop infringing activity may be less likely if an OSS license enforcement action is brought solely as a breach-of-contract claim rather than as an action for copyright infringement. For many years, injunctive relief was a presumptive remedy for copyright infringement; however, more and more courts are requiring the aggrieved copyright holder to show irreparable harm to obtain injunctive relief. Even so, the recipient should not presume that the royalty-free distribution of OSS software would, in and of itself, enable it to avoid an injunction. In *Jacobsen v. Katzer*,[33] discussed in more detail below, the Federal Circuit made it clear that under certain circumstances a copyright holder may be entitled to an injunction, even though the copyright owner released its software royalty-free to any party under an OSS license.[34] If a commercial product or business has been built around the use of certain OSS, an injunction preventing or restricting use of the OSS could have significant negative consequences for the success of such product or business.

Another potential form of equitable relief in an enforcement action is specific performance. As with injunctive relief, specific performance may be difficult to obtain because it generally requires a heightened showing of harm. However, if a court does order specific performance of an OSS license, there are some OSS licenses that have noteworthy terms. The GPLv3, for example, includes a provision that could, if enforceable as a contract term, require the OSS licensor to find a way to extend patent licenses that the OSS licensor has acquired from third parties to all downstream recipients.[35] Or consider this provision in the Jason Hunter license: "For a commercial project, permission is granted to use the com .oreilly.servlet.* packages provided that every person on the development team for that project owns a copy of the book Java Servlet Programming (O'Reilly) in its most recent edition."[36] The requirement to purchase a book may seem relatively trivial in contrast to obtaining the necessary rights from third parties to extend third-party patent licenses, which could be a difficult and expensive undertaking, but even a "trivial" noncompliance can be costly and embarrassing in the court of public opinion if not in a court of law.

As a consequence of these risks, it is important that recipients understand and comply with the terms of the licenses for the OSS they use. License compliance is not a trivial task for any entity that uses OSS because the licenses are often embedded in the OSS; there are no royalties to pay; there are no agreements to acknowledge, negotiate, or sign; and the license terms vary among the many licenses that are used. Entities cannot afford to be unaware that they are using OSS in their own products and services. The section below titled "Risk Mitigation Strategies" discusses policies and good practices for mitigating risks associated with OSS license compliance.

OSS Pedigree

Although OSS license compliance may be a paramount concern to recipients, code pedigree issues may be the more significant concern to the initial developers that desire to claim copyright ownership. Code pedigree refers to how well the ownership and intellectual property (IP) rights in the OSS are understood. OSS is very often developed in a community in which many contributors work collaboratively,[37] and having many authors complicates the chain of ownership. Much of the OSS discussed earlier in the section titled "OSS Licenses" is developed in these types of communities.

OSS code is said to have better pedigree when the source of each code contribution is not only known but documented, and the associated IP rights are assigned or licensed adequately to the party or parties that grant the licenses to the OSS. The party or parties claiming copyright ownership or control are often the individuals or organizations responsible for organizing or guiding the OSS development project. If contributors to an OSS project do not assign or license rights to one or more specific individuals or entities who in turn release the OSS, the ownership rights in the code being developed may be jeopardized. If no party (or set of parties) has clear ownership or exclusive license rights, it is unclear whether anyone would have standing to (1) enforce the copyright in the code against an infringer, or (2) enforce an OSS license against anyone breaching the OSS license or failing to comply with its terms.

In addition to jeopardizing the enforceability of an OSS license under which the code is released, poor pedigree may put a party using or releasing the OSS at risk of being an infringer. Without the proper assignment or license of rights, an OSS user or downstream distributor may be infringing the copyright rights associated with one or more contributions. Further, a contributor could submit code that he or she either does not own or lacks permission to contribute, e.g., an employee who submits his or her employer's code without permission, or an individual who submits code that is not his or her original work. In either case, a contribution without a license or an unauthorized contribution of another's work may taint that OSS code should the author or rightful owner of the code object to its use and distribution as part of the OSS project. Claims for IP infringement could be brought against the OSS licensor and the downstream recipients using the software. As such, those who undertake an OSS development project or develop or use OSS should carefully evaluate and mitigate risks involving the pedigree of the code they use. The section below titled "Risk Mitigation Strategies" discusses code integrity risk mitigation in more detail.

Patents and OSS

OSS licenses are typically focused primarily on copyright rights and issues because the text of software source code is protectable under copyright law. Software can also be protected by patents, for example, programmable apparatus, processes, or methods, yet many OSS licenses do not expressly provide a patent license to recipients. Those OSS licenses that do provide express patent licenses usually limit the patent license to the contributions the patent owner makes to the code base. For example, suppose someone downloads an OSS program (that person is now a recipient) and adds several files to that program (the recipient is now also a contributor and the OSS is now combined code). If the recipient/contributor releases that combined code under an OSS license that includes an express patent license limited to the contributor's contributions, the contributor would be licensing any patent rights the contributor might have to any downstream recipient to use the contributor's contributions, but not necessarily to use the OSS that was originally conveyed to the contributor, for free. The Mozilla Public License,[38] Eclipse Public License,[39] and Common Development and Distribution License[40] are OSS licenses that incorporate express patent licenses.[41] The BSD and MIT licenses (permissive) and the GPL and LGPL (restrictive) do not include any express patent licenses.

Without an express patent license, it is possible that an OSS license could be interpreted to include an implied patent license. Many OSS licenses are not precise in their licensing language and grant a license to "use" the software, which is an express patent right,[42] but not an express copyright right.[43] In addition, as an equitable matter, if a patent license is needed in order to exercise the grant of the copyright license, the patent license might be implied (at least with respect to all patents controlled by the OSS licensor). In other words, if an OSS license grants broad rights to copy and distribute software, it may be inequitable for the OSS licensor to assert patent rights that would prohibit or interfere with such copying and distribution. In this scenario, the recipient would defend a claim of patent infringement by arguing that the OSS licensor is legally estopped from asserting a patent that interferes with the express license to copy and distribute the OSS. As a result of imprecise legal language included in some OSS licenses and the possibility of derogating from the copyright license actually granted, patent licenses may be implied in some OSS licenses that do not expressly grant them.

Software users and licensors alike may also be concerned that the widespread use of OSS could be jeopardized by claims of patent infringement filed against the OSS. As a way of discouraging patent infringement claims, several OSS licenses have defensive termination provisions that terminate the patent license (and sometimes the copyright license) when a recipient asserts its patents. Depending on how the defensive termination provision is worded, termination may be invoked (1) if the recipient asserts its patents against the OSS licensor, (2) if the recipient asserts its patents against anyone using the OSS, or (3) if the recipient asserts its patents against either the OSS licensor or anyone using the OSS. Depending on the provision, termination may occur automatically or at the OSS licensor's option. Some termination provisions are triggered by the filing of a patent infringement lawsuit; whereas others are triggered by the mere threat of a patent infringement lawsuit.[44] Such defensive termination provisions will, however, affect only those that receive, use, or distribute the OSS and will not affect patent assertions by those parties that have no involvement in the development, use, or distribution of the OSS.

Each OSS license is different and, as a consequence, it is important for a recipient to review the specific licenses to understand (1) what patent rights, if any, are being licensed to recipients; (2) what patent rights, if any, a recipient is required to license to downstream recipients and the scope of such license if such recipients are further distributing the OSS; and (3) the possible ramifications if a recipient asserts a patent against a party on the basis of that party's use of OSS. The section below titled "Risk Mitigation Strategies" discusses risk mitigation strategies for those entities that maintain and license patent portfolios on a revenue-basis and could be substantially affected by the terms of OSS licenses.

OSS Litigation

Historically, there has been relatively little litigation involving OSS, but it appears to be increasing. In particular, the battle among mobile operating platforms has brought an onslaught of patent litigation directed at the Android platform—an OSS platform developed by Google. Several handset manufacturers have incorporated Android into their handsets as an increasing number of applications developers release applications written for Android. Their inclusion of the Android software, which is distributed under an OSS license, has formed the basis for many patent infringement claims between handset manufacturers.[45] This illustrates the previous point that OSS software licensed royalty-free is not necessarily free from patent infringement. A recipient may be in full compliance with the OSS license but still be at risk of an infringement claim. Moreover, unlike commercial software licenses where it is fairly common for the owner of the software to agree to defend and indemnify its licensees if they are sued for patent infringement by a third party,

OSS licensees do not typically receive similar protections without a separate commercial contract.

OSS may also be the basis for copyright infringement claims. One Android case, notable for its copyright pedigree issues, involves claims by Oracle America, Inc. (Oracle) that certain Java programs included in Android infringe Oracle's copyrights. One aspect of the lawsuit deals with which elements of a program or software interface are copyrightable expressions. Oracle asserts that the Android code is directly copied from copyrighted Oracle code and that about one-third of Android's application programming interfaces (APIs) are derived from Oracle's copyrighted APIs and documents.[46] In an interim order, the court found that the names of Java language API files, packages, classes, and methods are not protected because words and short phrases are not copyrightable. In a subsequent order, the court ruled that the replicated elements of the structure, sequence, and organization of the Java APIs are not protected by copyright because the command structure is a system or method of operation under Section 102(b) of the Copyright Act.[47] The court emphasized that its holding regarding the copyrightability of the structure, sequence, and organization of computer programs is specific to the facts of this case.[48]

Although sparse, a few lawsuits have dealt with license compliance issues. In *Jacobsen v. Katzer*,[49] Jacobson sued Katzer for failing to comply with the terms of the Artistic license. The Artistic license is a permissive OSS license that requires a recipient to reproduce the copyright notices and disclaimers and provide descriptions of how the files or computer code have been modified from the original source code when the recipient redistributes the OSS to third parties.[50] Katzer failed to fulfill these requirements, and the Federal Circuit held that the terms of the Artistic License are enforceable copyright conditions.[51] Importantly, the *Jacobsen* case illustrates that a user of OSS that violates the conditions of that OSS license may be a copyright infringer and may not be merely in breach of an agreement. Moreover, the Federal Circuit determined that infringement of OSS could warrant an injunction.[52]

In recent years, the Software Freedom Law Center has filed a series of complaints on behalf of the copyright holders of the BusyBox software.[53] BusyBox is a set of UNIX utilities for limited resource devices that is distributed under the GPLv2.[54] The GPLv2 is one of the most restrictive OSS licenses in that distributors of the GPL software must reproduce the GPL with each copy they distribute and make the source code available (or at least notify recipients that they can request a copy of the source code). The litigation surrounding BusyBox revolves around a common theme: manufacturers integrated BusyBox into their products and failed to provide access to source code as required under the license. The lawsuits typically settled after agreement by the offending company to distribute the source code, notify past recipients, appoint an open source compliance officer, and pay financial consideration to the developers of BusyBox.[55] Because these types of cases have tended to settle and not be decided in a court of law, OSS licensors and licensees do not yet know what courts will decide, e.g., with respect to specific performance to release of source code, imposition of injunctions, and calculation of damages.[56]

RISK MITIGATION STRATEGIES

Just as there are many OSS licenses and OSS development projects, there are many strategies that may be applied to mitigate the risks discussed in the previous section titled "Legal Risks Associated with OSS." Any strategy adopted must necessarily account for the manner and extent to which OSS is being used by an entity, the nature and culture of the entity, and the entity's objectives and risks associated with its use of OSS. For example, an individual programmer who incorporates OSS into a program the programmer is developing as a hobby requires a very different risk mitigation strategy from a large enterprise with hundreds of employees developing commercial software that may incorporate millions of lines of OSS code.

Individual Developers

Most individuals who are developing free programs that are unlikely to compete with any substantial commercial undertaking are not likely targets of lawsuits seeking damages, injunctions, or specific performance of various terms that may be present in an OSS. Notwithstanding this unlikelihood, the *Jacobsen* case discussed above represents an example of a significant dispute and protracted legal battle that occurred essentially between individuals and involved software for model railroads. Individual developers should therefore ensure that they comply with the license terms governing the software they choose to include in the programs they develop. When software is downloaded, it is important to look for the license that spells out how the downloaded software may be used, whether it may be copied, distributed, or modified, and if so, what the recipient must do to comply with the license granting these rights. The following are a few good practices individual developers, as well as those developing software as employees, might follow to reduce common noncompliance risks:

- If downloading source code, search the OSS source code comments for any copyright or license language.
- Search the text files, such as Readme, Notice, and Legal files, downloaded with OSS executable code for any copyright or license language.
- If incorporating downloaded OSS into a program, include the copyright and license language (including any disclaimers) in each source file, or a separate text file when the source code is compiled or when the code is available only in executable form.
- If the downloaded OSS is subsequently modified, follow any instructions for noting such modifications according to the terms of the applicable license.
- If incorporating downloaded OSS into a program, ensure that all other license terms are followed such as making the source code available or providing an offer to make the source code available.
- If the OSS license covering the downloaded code applies to any modifications to the downloaded code or other code that is combined with the downloaded code, ensure that any such modifications or combined code is made clearly available under the same OSS license.

It is often difficult to understand how a particular OSS license may affect a proprietary program when the OSS subject to that OSS license is combined or integrated with a proprietary program. Individual programmers should, of course, seek legal advice from counsel experienced with OSS licensing matters. If seeking such advice is untenable, however, a developer should consider asking the copyright holder or licensor of the OSS whether or not the developer's planned use or distribution of the OSS complies with the terms of the OSS license.

Organizations

Any organization, large or small, for-profit or nonprofit, should consider implementing an OSS internal policy that employees, contractors, and other representatives of the organization should follow when developing and distributing the organization's software. The details and requirements of such an internal policy should vary from organization to organization, and each policy works best when it is carefully crafted with the particular organization and its activities in mind.[57] In general it is a good practice to train all of the organization's developers to follow the practices listed above under the heading "Individual Developers."

In addition, the organization might consider including guidelines or requirements in its policy addressing the following issues:

- Who within an organization decides whether or not OSS may be included in the organization's programs? At what point in the development prior to release is that decision made?
- Who is responsible for making sure that all OSS licenses have been properly complied with?
- Is all OSS incorporated into the organization's programs tracked, and if so, by whom?
- Is the use of OSS subject to certain OSS licenses always permitted? Always denied unless there is an exception after an internal escalation?
- Is the use of OSS developed within certain projects always permitted? Always denied unless there is an exception after an internal escalation?
- Are certain uses permitted for internal use only, and if so, who is required to ensure that the use is only internal?
- When, if at all, may an employee or other representative of the organization participate in a community-based OSS development project?

Communications and training cannot be overemphasized. One key focus of an organization that adopts such a policy is communicating the policy to its employees and representatives that are expected to follow the policy. Often OSS training is required to educate relevant employees and representatives about the risks associated with the use of OSS in an organization's products and services and to ensure that the organization's policy is known and understood by such employees and other representatives.

Entities should also consider how they should respond to a claim that they failed to comply with an OSS license. Based on the cases discussed above, prompt compliance tends to significantly reduce the likelihood of protracted litigation. Nonetheless, demands for compliance that might require the release of proprietary source code may not be desirable. In such situations, constructive and respectful discussions with the owner of the OSS may be helpful. Keep in mind that OSS licensors are rarely interested in licensing their software for the purpose of causing another developer to inadvertently license proprietary software in ways that defeat the commercial viability of such proprietary software. For this reason, it may also be helpful for an entity, especially larger entities, to ensure that they have internal processes in place to receive OSS compliance questions and to route such questions promptly to experts within the entity who can properly investigate any OSS-related complaints or questions, authorize remedial actions, or contact the OSS copyright licensor to discuss acceptable options, as may be appropriate.

CONCLUSION

As the field of bioinformatics continues to grow, so will the use of OSS in biological and medical sciences. Individual software developers and entities that develop, distribute, or use bioinformatics software should understand the license requirements for the OSS they incorporate into such bioinformatics software as well as the pedigree of the OSS they use. Appropriate risk mitigation strategies may be employed to reduce the likelihood of violating intellectual property rights associated with the use of OSS or failing to comply with relevant OSS licenses.

NOTES

1. Stephen Friend, *Opinion: Thinking Outside the Genome*, SCIENTIST, MAG. LIFE SCI. (Oct. 1, 2011), http://the-scientist.com/2011/10/01/opinion-thinking-outside-the-genome/.

2. *Open Source Initiative*, THE OPEN SOURCE DEFINITION, http://www.opensource.org./docs/osd (last visited Jan. 6, 2012).

3. A party may interact with OSS in a variety of ways. For purposes of this chapter, we use the following terminology: "Initial developer" means the original party that releases OSS under an OSS license; "recipient" means a party receiving OSS; "conveyed OSS" means the OSS initially received by a recipient; "OSS licensor" means a party that conveys OSS, whether or not that party has modified the OSS; "downstream recipient" means any party that receives the OSS along the chain of distribution; "contributor" means a recipient that adds code to OSS code, which the contributor then distributes; "combined code" means code developed by a contributor combined with conveyed OSS.

4. OSI's OSD states the requirement as follows: "The rights attached to the program must apply to all to whom the program is redistributed without the need for execution of an additional license by those parties." *Open Source Initiative*, THE OPEN SOURCE DEFINITION § 7, http://www.opensource.org./docs/osd (last visited Jan. 6, 2012).

5. For example, OSI has approved approximately 70 separate OSS licenses. *Open Source Initiative*, LICENSE BY NAME, http://www.opensource.org/licenses/alphabetical (last visited Jan. 6, 2012).

6. The GPLv3 license, however, allows for limited additional terms. FREE SOFTWARE FOUNDATION, INC., GNU GENERAL PUBLIC LICENSE VERSION 3 § 7 (2007).

7. While there would be little incentive to accept a commercial license restricting access to source code or imposing fees on the use and distribution of the software when such software is available under an OSS license, many commercial licenses provide other terms and protections for recipients that may make such tradeoffs prudent. For example, some commercial license agreements also provide for service, support, and/or other related products and services. Commercial agreements may also include representations, warranties, and/or indemnities regarding the software's performance or pedigree. Dual-licensing approaches are discussed in "Increasing Importance of Bioinformatics Open Source Software."

8. *Open Source Initiative*, THE BSD 3-CLAUSE LICENSE, http://www.opensource.org/licenses/BSD-3-Clause.

9. *Open Source Initiative*, ARTISTIC LICENSE 2.0, http://www.opensource.org/licenses/Artistic-2.0.

10. *Open Source Initiative*, THE MIT LICENSE (MIT), http://www.opensource.org/licenses/MIT (last visited Jan. 6, 2012).

11. *Id.*

12. *GNU Operating System*, LICENSES, http://www.gnu.org/licenses/licenses.html.

13. *Free Software Foundation*, GNU OPERATING SYSTEM, LICENSES, http://www.gnu.org/licenses/gpl.html (last visited Jan. 6, 2012).

14. John Quackenbush, *Open-Source Software Accelerates Bioinformatics*, 4 GENOME BIOLOGY 336, 336.3 (2003), *available at* http://genomebiology.com/content/pdf/gb-2003-4-9-336.pdf (noting that the approach of researchers carefully guarding their software source code "is rapidly fading away and being replaced by a commitment among developers to create open-source software tools that can be used, adapted, and improved by the wider scientific community").

15. Jason E. Stajich & Hilmar Lapp, *Open Source Tools and Toolkits for Bioinformatics: Significance and Where Are We?*, 7 BRIEFINGS IN BIOINFORMATICS 287, 287 (2006); Sandrine Dudoit, Robert C. Gentleman & John Quackenbush, *Open Source Software for the Analysis of Microarray Data*, 34 BIOTECHNIQUES 45, 46 (2003).

16. Stajich & Lapp, *supra* note 15.

17. *Id.*

18. *Bioinformatics.org*, FULL GROUP LISTING, http://www.bioinformatics.org/search/fullprojectlist.php.

19. University of North Carolina Center for Bioinformatics, *Open Source Tools*, http://bioinformatics.unc.edu/software/opensource/index.htm.

20. OPEN BIOINFORMATICS FOUNDATION, http://www.open-bio.org/wiki/Main_Page.

21. Microsoft Research, *.NET Bio*, http://research.microsoft.com/en-us/projects/bio/mbf.aspx.

22. *Microsoft Biology Foundation Evolves into New Toolkit: .NET Bio*, MICROSOFT RESEARCH CONNECTIONS BLOG, http://blogs.msdn.com/b/msr_er/archive/2011/10/18/microsoft-biology-foundation-evolves-into-new-toolkit-net-bio.aspx.

23. Mark A. Lemley & Ziv Shafir, *Symposium: The Licensing of Intellectual Property: Who Chooses Open Source?*, 78 U. Chi. L. Rev. 139, 141–42 (2009).

24. *Id.* at 141; *see also* Affymetrix, http://www.affymetrix.com.

25. Bioclipse, http://bioclipse.net/.

26. *OpenClinica*, Product Editions Overview, https://www.openclinica.com/product-editions-overview.

27. *Open Source Initiative*, Artistic License 2.0, http://opensource.org/licenses/Artistic-2.0.

28. *Open Source Initiative*, The BSD 3-Clause License, http://www.opensource.org/licenses/BSD-3-Clause.

29. *The Apache Software Foundation*, Licenses, http://www.apache.org/licenses/.

30. 17 U.S.C. § 504(c)(1).

31. *Id.* § 504(c)(2). The court may also award reasonable attorneys' fees. *Id.* § 505.

32. *See, e.g.*, Tattoo Art, Inc. v. TAT Int'l., LLC, 794 F. Supp. 2d. 634 (E.D. Va. 2011) (stating "the copyrights in multiple works may be registered on a single form, and thus considered one work for the purposes of registration while still qualifying as separate 'works' *for purposes of awarding statutory damages*") (emphasis added); Oracle Am., Inc. v. Google Inc., No. C 10-03561 at 6 (N.D. Cal. Sept. 15, 2011) (order partially granting and partially denying defendant's motion for summary judgment on copyright claim) ("when a single published unit contains multiple elements 'that are otherwise recognizable as self-contained works,' the unit is considered a single work for the limited purpose of registration, *while its elements may be recognized as separate works for other purposes*.") (emphasis added).

33. Jacobsen v. Katzer, 535 F.3d 1373 (Fed. Cir. 2008).

34. *Id.* at 1381–82 ("Copyright licenses are designed to support the right to exclude; money damages alone do not support or enforce that right. The choice to exact consideration in the form of compliance with the open source requirements of disclosure and explanation of changes, rather than as a dollar-denominated fee, is entitled to no less legal recognition. Indeed, because a calculation of damages is inherently speculative, these types of license restrictions might well be rendered meaningless absent the ability to enforce through injunctive relief.").

35. Free Software Foundation, GNU General Public License Version 3 § 11 (2007), http://www.gnu.org/licenses/gpl.html ("If you convey a covered work, knowingly relying on a patent license, and the Corresponding Source of the work is not available for anyone to copy, free of charge and under the terms of this License, through a publicly available network server or other readily accessible means, then you must either (1) cause the Corresponding Source to be so available, or (2) arrange to deprive yourself of the benefit of the patent license for this particular work, or (3) *arrange, in a manner consistent with the requirements of this License, to extend the patent license to downstream recipients*.") (emphasis added).

36. Olex OpenLogic Exchange, Jason Hunter License (2002), http://olex.openlogic.com/licenses/jasonhunter-license.

37. Such community development is not a requirement of OSS but rather a common attribute of OSS.

38. Mozilla, Mozilla Public License Version 1.1 §§ 1.1, 2(b), *available at* http://www.mozilla.org/MPL/MPL-1.1.html.

39. Eclipse, Eclipse Public License Version 1.0 §§ 1, 2(b), *available at* http://www.eclipse.org/legal/epl-v10.html.

40. Open Source Initiative, Common Development and Distribution License (CDDL), Version 1.0 § 2.2, *available at* http://www.opensource.org/licenses/cddl1.php.

41. For an in-depth discussion of OSS licenses with express patent grants, see Michele Herman & Joanne Montague, *The Elephant in the Room: Patent Value and Open Source Software* § II.B (AIPLA 2011 Spring Meeting Speaker Materials) (2011), *available at* http://www.dwt.com/portalresource/lookup/wosid/intelliun-1501-42003/media.name=/Montague%20-%20AIPLA%202011%20Spring%20Meeting%20Paper.pdf.

42. 35 U.S.C. § 271(a) states: "Except as otherwise provided in this title, whoever without authority makes, *uses*, offers to sell, or sells any patented invention, within the United States, or imports into the United States any patented invention during the term of the patent therefor, infringes the patent." (emphasis added).

43. 17 U.S.C. § 106. Copyright rights include the rights to reproduce, prepare derivative works of, distribute copies of, perform, and display the copyrighted work. For an in-depth discussion of OSS licenses without express patent grants, see Herman & Montague, *supra* note 41, at § II.A.

44. There are at least three general ways to categorize these defensive termination provisions: (1) the recipient loses its OSS license (copyrights and patents) if the recipient claims that the OSS infringes the recipient's patent; (2) the recipient loses the patent licenses granted through the OSS license if the recipient claims that an OSS licensor is infringing the recipient's patent; and (3) the recipient loses its OSS license (copyrights and patents) if the recipient claims that the OSS licensor infringes a recipient's patent. Herman & Montague, *supra* note 41, at § IV.

45. Defendant Apple Inc.'s Answer, Affirmative Defenses, and Counterclaims to Plaintiff Motorola Mobility, Inc.'s First Amended Complaint for Patent Infringement, Motorola Mobility, Inc. v. Apple Inc., No. 1:12-cv-20271 (S.D. Fla. June 4, 2012); Complaint for Patent Infringement, Apple Inc. v. High Tech Computer Corp., No. 1:10-cv-00544 (D. Del. June 21, 2010); Complaint of Apple Inc. et al. Under Section 337 of the Tariff Act of 1930, as amended, *in re* Certain Personal Data and Mobile Communications Devices and Related Software, No. 337-TA-710 (U.S.I.T.C. Mar. 2, 2010).

46. Amended Complaint ¶ 40, Oracle Am., Inc. v. Google, Inc., No. 3:10-cv-03561 (N.D. Cal. Oct. 27, 2010).
47. Oracle Am., Inc. v. Google Inc., 872 F. Supp. 2d 974, 976–77 (N.D. Cal. 2012).
48. *Id.* at 1002.
49. Jacobsen v. Katzer, 535 F.3d 1373 (Fed. Cir. 2008).
50. *Id.* at 1376.
51. *Id.* at 1382–83.
52. *Id.* at 1383 ("The choice to exact consideration in the form of compliance with the open source requirements of disclosure and explanation of changes, rather than as a dollar-denominated fee, is entitled to no less legal recognition. Indeed, because a calculation of damages is inherently speculative, these types of license restrictions might well be rendered meaningless absent the ability to enforce through injunctive relief.").
53. Press Release, Software Freedom Law Center, Best Buy, Samsung, Westinghouse, and Eleven Other Brands Named in SFLC Lawsuit (Dec. 14, 2009), http://www.softwarefreedom.org/news/2009/dec/14/busybox-gpl-lawsuit/; Press Release, Software Freedom Law Center, SFLC Files GPL Violation Lawsuit Against Extreme Networks, Inc. (July 21, 2008), http://www.softwarefreedom.org/news/2008/jul/21/busybox/; Press Release, Software Freedom Law Center, SFLC Files Another Round of GPL Violation Lawsuits on Behalf of BusyBox Developers (June 10, 2008), http://www.softwarefreedom.org/news/2008/jun/10/busybox/; Press Release, Software Freedom Law Center, BusyBox Developers File GPL Infringement Lawsuit Against Verizon Communications (Dec. 7, 2007), http://www.softwarefreedom.org/news/2007/dec/07/busybox/; Press Release, Software Freedom Law Center, Second Round of GPL Infringement Lawsuits Filed on Behalf of BusyBox Developers (Nov. 20, 2007), http://www.softwarefreedom.org/news/2007/nov/20/busybox/; Press Release, Software Freedom Law Center, On Behalf of BusyBox Developers, SFLC Files First Ever U.S. GPL Violation Lawsuit (Sept. 20, 2007), http://www.softwarefreedom.org/news/2007/sep/20/busybox/.
54. *BusyBox*, BoyBox: The Swiss Army Knife of Embedded Linux, http://busybox.net/about.html (last visited Jan. 8, 2012).
55. Press Release, Software Freedom Law Center, BusyBox Developers Settle Case with Extreme Networks (Oct. 6, 2008), http://www.softwarefreedom.org/news/2008/oct/06/busybox-extreme-settle/; Press Release, Software Freedom Law Center, BusyBox Developers and Supermicro Agree to End GPL Lawsuit (July 23, 2008), http://www.softwarefreedom.org/news/2008/jul/23/busybox-supermicro/; Press Release, Software Freedom Law Center, BusyBox Developers Agree to End GPL Lawsuit Against Verizon (Mar. 17, 2008), http://www.softwarefreedom.org/news/2008/mar/17/busybox-verizon/; Press Release, Software Freedom Law Center, BusyBox Developers and High-Gain Antennas Agree to Dismiss GPL Lawsuit (Mar. 6, 2008), http://www.softwarefreedom.org/news/2008/mar/06/busybox-hga/; Press Release, Software Freedom Law Center, BusyBox Developers and Xterasys Corporation Agree to Settle GPL Lawsuit (Dec. 17, 2007), http://www.softwarefreedom.org/news/2007/dec/17/busybox-xterasys-settlement/; Press Release, Software Freedom Law Center, BusyBox Developers and Monsoon Multimedia Agree to Dismiss GPL Lawsuit (Oct. 30, 2007), http://www.softwarefreedom.org/news/2007/oct/30/busybox-monsoon-settlement/.
56. Willful infringement was found by one court in a default judgment; the court granted a permanent injunction, awarded treble statutory damages, required forfeiture of infringing articles, and stated that the plaintiffs are entitled to reasonable attorneys' fees. Software Freedom Conservancy v. Best Buy Co., 2010 U.S. Dist. LEXIS 75208, at *10, *12, *15, *17 (S.D.N.Y. 2010).
57. *See* Creating an Open Source Policy for Your Company, Intellectual Property Owners (IPO) Association Open Source Committee, draft Oct. 29, 2004, http://www.ipo.org/AM/Template.cfm?Section=Search&Template=/Search/SearchDisplay.cfm.

CHAPTER 4

Software Development

Heather J. Meeker

WHAT'S NEW TODAY?

Before I went to law school, I worked as a paralegal. My supervisor was the vice president of legal affairs for a large company. When I was contemplating law school, I asked him what it was like to be a lawyer. He told me that in the 15 years he had been at the company, every day he came into the office, he dealt with something new, and that was why he loved the practice of law. Today, I have been handling software deals for more than 15 years, and I understand exactly what he meant.

In one sense, software deals are software deals, and they are conducted by software lawyers, and they are the same no matter what the software does. In a sense, developing software, and stewarding the intellectual property (IP) it embodies, is similar from one software application to the next. Software that presents and manipulates bioinformatics in graphic format is similar to computer-aided design (CAD) software; data consisting of strings of proteins is stored in the same kind of bits and bytes as financial data. Plenty has been written about how to conduct software deals, and there is no need to reinvent that wheel. This chapter focuses instead on how bioinformatics deals are different from other kinds of software deals. It provides a road map for the general software transactions practitioner to ease into the water of bioinformatics.

This chapter discusses inbound licensing and development deals interchangeably. That is because almost nothing is developed from scratch today. Particularly given the advent of open source software, which provides ubiquitous tools and building blocks free of charge, almost all software licenses are partially development deals, and almost all software development deals are partially licenses.

As a baseline, though, we should understand the issues that arise most frequently in all software development deals. They are:

1. IP ownership
2. IP indemnities
3. Testing and acceptance (and payment)
4. Open source and third-party software

IP Ownership

One of the hardest tasks in negotiating development or other inbound licensing transactions is determining who owns the results. Of course, software developers today usually enter a development project with their own tricks and tools, and they are loath to give up the right to use them for a single customer, so simply assigning over rights in all deliverables is almost never viable now. Conversely, customers who pay for a standard or existing product to be modified are disinclined to share with the world the improvements developed on their own nickel. But the most taxing and frustrating aspect of negotiating ownership provisions is that one is always trying to claim rights in what does not exist, and in fact may not ever exist, and in what as a baseline is likely to be ill-defined. Transactional lawyers address this by creating "buckets"—definitions of terms like foreground intellectual property, background intellectual property, improvements, preexisting intellectual property, and so forth. Defining the buckets is hard; allocating buckets is not so hard. But the allocation depends on the price, of course. Most customers are willing to trade ownership for discounts, and as a business and cultural matter, people expect to own what they pay for.

IP Indemnities

If I hand you some software, you probably want to be sure I haven't stolen it from someone else. That is the easy part of indemnities; most licensors or developers will undertake liability for claims of copyright infringement or trade secret infringement that they have introduced via the development process. Patents are much trickier, because one can infringe a patent without knowledge or ill intent. Developers and licensors are therefore hesitant to undertake liability for unknown patent infringement—particularly in this age of nonpracticing entities (NPEs). All this means that the indemnities, warranties, and limitations of liability in a software development license agreement can be complex and contentious. In software agreements, these issues are worked through in a sometimes complex combination of warranties, indemnities, and liability limitations.

Testing and Acceptance (and Refunds)

Any deal involving development of custom software—whether it consists of customizations of an existing product, integration of existing software in a particular platform or environment, or new software built from scratch—is risky for both the developer and customer. Developers must adapt for the dreaded "scope creep," and customers must do their best to approve specifications for work they don't usually understand. Most software agreements, therefore, include some kind of "performance warranties" that the software will work according to specifications. If not, the customer gets a refund.

But in reality, it is not that simple. Customers and developers usually share the risk for a failed development. Attending this issue is a quirk of the U.S. accounting rules, which will not allow a developer to recognize revenue on fees unless there is effectively no likelihood that a refund will occur. If the deal involves developing software from scratch, this is pure business risk. But if a portion of the fees is for a license to preexisting software, the developer will expect to realize that revenue without contingency. This means that developers seeking to make their accounting quarters will be very resistant to including refund provisions in agreements. This is one of the single most contentious issues in software development contracts.

Open Source and Third-Party Software

This issue is not new, but the increasing reliance on open source software components has brought it to the forefront. Any time a developer includes third-party software in a deliverable, there is the possibility that the software will not be properly licensed, and the customer's use of the software will infringe the rights of the third party that developed the software. In contemporary software development, developers rely heavily on open source libraries and components. Because these components are provided under licensing terms that are sometimes challenging to interpret, and because those terms disclaim all warranties and liabilities on the part of the third-party developer, customers often take a skeptical view of a developer's decision to include third-party components. Whereas formerly a customer might have been satisfied with an indemnity covering intellectual property claims, customers today often require detailed disclosures regarding the use of open source software in deliverables, and raise serious "due diligence" questions about the interpretation of open source licenses—questions that must be answered by the developer before the customers will sign the contract or accept the software.

One final word about patent infringement liability. Today's customers are increasingly concerned about software patent infringement liability. Although this is a serious concern, it is not particular to bioinformatics or to open source software. Developers and licensors have always been resistant—and for good reason—to undertaking liability for third-party patent infringement for custom development. A seller of an off-the-shelf software product might be reasonably expected to undertake liability for patent infringement, because that liability is an unavoidable cost of its business. But custom development could be infringing of third-party patents without either the developer or the customer being aware of the liability. Allocating that risk is always contentious in any software license or development agreement.

Other Issues

Other issues that typically arise in software development agreements include:

(a) Basis of payment: time and materials, deliverable/acceptance based, or hybrid (success fees or late penalties)
(b) Contingencies: customer obligations on which development activities depend
(c) Third-party licenses: in addition to issues with third-party technology embedded in deliverables, the use of deliverables after delivery may require licenses to third-party platforms or environments such as operating systems and language platforms.
(d) Updates: keeping deliverables current and error-free
(e) Escrows: for software provided in object code form only, escrows of source code to be accessible upon support failure, business distress, or other triggers

Plenty has been written on these topics outside the bioinformatics context, so there is no need to belabor them here. But they will also come up in any bioinformatics deal, and will in some cases color the treatment of the more specific bioinformatics issues outlined below.

THE CURIOUS CASE OF THE BIOINFORMATICS DEAL

Now we get to the interesting part. When I arrive at work in the morning with a bioinformatics deal on my desk, what is new on that day? There are, I think, five primary issues that arise particularly from software development in bioinformatics.

Data by the Terabyte—High-Performance and Nonstandard Platforms

Bioinformatics software handles a torrent of data: both as input, which can include large and complex databases representing genetic information, and as output, which can include sophisticated computer graphics. In addition, because the format of bioinformatics data is not yet standardized, significant processing may be necessary to convert or import data from one format to another.

These tasks, from a computing standpoint, are different from the day-to-day business processing for which most businesses use computers. They require more computing power and speed, and high-performance access to storage. While, in a sense, all computing tasks involve the movement of bits and bytes, scale does make a difference. The scale of bioinformatic processing means that the computers, storage, and software utilities that you buy at Office Depot probably won't run bioinformatics software. Bioinformatics software also tends to be as idiosyncratic as bioinformatics projects themselves. Standard software packages can require tweaking to fit the data set or analysis technique, and those tweaks can turn a standard, easily supportable product into a unique, unsupportable one.

This raises a set of issues that has, to some degree, fallen away from most contemporary computing deals. In most software licenses today, compatibility is handled easily by specifying that the software will be compatible with the latest versions of Windows, Linux, or one of the mobile platforms. Twenty or thirty years ago, that was not the case: all software development was for a specific and often unique computing environment. Today, bioinformatics software may need to be compatible with nonstandard operating system platforms, or in turn, enhanced platforms may be necessary to run the bioinformatics software. High-volume number crunching may also require enhanced data storage and retrieval systems. Complex graphics may require ultra-fast processors or graphics utility software. Even if you are operating in a "standard" Linux system—which is particularly common for academic projects—your environment may be heavily customized.

When preparing bioinformatics development or licensing deals, special care should be taken to avoid bugs and incompatibilities arising from nonstandard computing environments. Moreover, each side should consider the likely shelf life of a nonstandard platform. If the platform goes through an end-of-life cycle, the software running on it may no longer function. Most licensees or customers expect that their software will run forever, and though most are willing to commit to install the next update of a standard system like Microsoft Windows, moving to the next version of a nonstandard system may be expensive, difficult, or impossible. In a bioinformatics deal, this means customers will require commitments to port or update software to meet changing platform requirements. On the developer end, it may be difficult to staff development or customization work for nonstandard platforms, so developers will be understandably resistant to long-term commitments.

One—perhaps unexpected—way in which this may play out in a development contract is in employee nonsolicitation provisions. An engineer with the skills and the know-how to work in a custom environment may be very valuable to both the customer and the developer. Most software development contracts contain some kind of provision preventing the customer from "poaching" the developer's employees, but in bioinformatics deals, a smart customer will vigorously resist agreeing to such restrictions.

A Picture Is Worth a Thousand Amino Acids—Graphics Formats

In today's computing, much is made of pretty user interfaces—and rightly so, because user interfaces are key to product quality. But in truth, even today, the great majority of program interfaces consist entirely of text, formatting, embedded video, and static images. Software interfaces today are still mostly static. However, some software makes heavy use of computer graphics, and software that does so is a different breed.

Any sensible developer in this area stands on the shoulders of giants. There is an array of 3-D graphics formats in common use, and these formats all come with their own business and intellectual property idiosyncrasies.

Fortunately, graphics formats and the codecs that implement them have not suffered from the same patent wars that have broken out for audio and motion video formats. Unfortunately, the lack of controversies means the information about patent coverage for graphics codecs is harder to come by than in the more litigious fields. In fact, there has been no patent litigation among the major players in this area, probably because they all write codecs for each other's formats. A notable exception was a case that dealt with the trademark protection of a CAD format in which it was determined that a three-letter file extension for a format did not merit trademark protection.[1]

If your bioinformatics deal involves using third-party graphics formats, you should understand the intellectual property landscape behind your choice. First, it is key to understand that a format and a codec are two very different things. Formats are like formulae for how graphics will be represented in digital processing. They are primarily protected by patent—if their use is restricted at all. A codec, in contrast, is a particular implementation of software that reads and writes the format. Clearly, there can be many codecs for any given format, just as there might be many players that play a CD or DVD. Those codecs are software, and are primarily governed by copyright. They may be open source or proprietary, for instance, and they may be free of charge or require licensing fees. Many of the companies that promulgate file formats provide codecs or tools that read and write them. The IP issues for a given format will depend greatly on the codecs being used to process them.

The bottom line is that graphics processing software will be part of most bioinformatics deals, and that customers and developers alike should understand that IP clearance for the format and the codec being used should be considered separately.

The Anarchic World of Software—Regulation

To me, one of the joys of working in the software field is that it is like the Wild West. Software development contracts are a law unto themselves; almost none of their parts is heavily dictated by regulation. This means that the parties are greatly in control of their relationship, and those who draft and negotiate deals have wide latitude to do business as they please. Contrast that with, say, an insurance contract—a relationship in a highly regulated field where even specific language of contract provisions may be required by law.

Information in life sciences, however, can be highly regulated. Any personally identifiable information will be subject to significant regulation in the United States and also, even more so in Europe. It may be that your data cannot be transmitted across national boundaries without adhering to regulatory concerns.

In most software development contracts, little attention is paid to the data being processed. Contracts for bioinformatics development or processing, however, should contain not only customary confidentiality and nondisclosure provisions but requirements to adhere to applicable health and safety law. Also, there may be restrictions on exporting certain sensitive information, under relevant U.S. Department of Commerce or Department of Defense rules.

One particular case in which data transfer can run afoul of the law unexpectedly is where a user contracts for support for software. If the user finds a bug, the party providing support will usually want to test the software on the exact data that initially generated the bug. But if the party using the software and the party providing support are in different countries, that can raise export concerns—a frequent situation in this age of outsourced development. Also, secrecy and privacy of data can be compromised if live data is transferred to development environments, which often do not have the same protocols for data control as live environments.

Finally, bioinformatics, unlike most software applications, often requires the extensive use of third-party databases. Just as software developers are customarily required to clear rights to embedded third-party software components, they may be expected to secure rights to access databases. The bioinformatics contract should specify which party will clear access to any third-party databases, and which will have responsibility for any fees or regulatory requirements attendant to that access.

I Play for the Users—Testing and Acceptance

Testing and acceptance terms are always some of the most difficult terms to negotiate in a software development or license agreement. This is because each side will have significant competing interests in the deal. A user or licensee wants to be sure it is not paying for software that does not work—either because the software has bugs, or because it simply does not work for the environment in which it is planned for use, as covered in the custom environments discussion previously. A developer or licensor does not want to engage in significant work for no compensation on a project where the user has incorrectly specified its needs, or changes its requirements midstream so that what "works" becomes a moving target.

The reality is that all development projects are moving targets. No significant software development project in the history of the world has ever been specified properly at the outset. Specifications are often living documents. The more customized the project, the bigger the problem. To avoid the negotiation becoming a tug-of-war, each party should recognize this likelihood at the outset. The initial specifications for a development must often include interim deliverables that consist of specifications or acceptance criteria for the specification details that cannot be determined at the outset. Development agreements should include processes for change orders and updates to the project plan, as well as detailed acceptance testing criteria. When you are structuring or drafting this type of agreement, remember that process is the next best thing to an answer. If you cannot solve an issue when negotiating the contract, at the least you can determine a sensible process for sorting out the issue as you go forward.

However, no amount of process will dispose of the ultimate risk of project failure. This risk devolves to a question of payment terms. At the two ends of the spectrum are payment on a pure time-and-materials basis—which favors the developer—and a pure milestone/acceptance basis—which favors the customer. For an entirely custom project, or one that has many technological moving parts (including subcontracting or multiple contractors), time and materials may be appropriate. In that case, the customer will be in the same risk position as if it were using internal resources to complete the project; no business risk is transferred to the developer. For a project that consists of minor customization to existing software, or perhaps only professional services to knit together different programs into a particular environment, and significant licensing fees for the existing parts of the solution, a pure milestone payment scheme may be appropriate.

But there are shades of gray in between. Many development agreements contain hybrid payment terms that combine elements of both approaches. For instance, time and materials arrangements may contain penalties for late delivery, or bonuses for on-time or underbudget completion, or not-to-exceed amounts. For a complex development project, it may be useful to conduct the development in phases that can be triggered at the customer's discretion. A first phase could include initial specifications or project plans. Such a phase is usually not expensive, and allows the customer to take a smaller risk to get more information to assess ongoing risk. At the end of such an initial phase, the customer may wish to put the project out for bid among various developers. As the project proceeds, the customer will usually bear less risk in the deal terms, with the final phase being a pure milestone/acceptance payment—but at that point, the risk of project failure will be significantly reduced for both sides.

It is important to understand that almost no amount of contract verbiage can ensure project success—or even perfect clarity as to where fault might reside for failure. Any project of meaningful size will deviate quickly from the approved statement of work, as customer and developer both make adjustments to move the project forward. When a project goes wrong, and a lawyer asks a client to reconstruct what was actually done, and what changes were made and approved, the written evidence is always chaotic and inconsistent. In a perfect world, all changes are described in change orders, and a person with proper authority signs all approvals and acceptances. But in the real world, reconstructing the facts of a development dispute can be very costly. What is important is the relationship; the contract is just a road map to finding the debris if the relationship goes bad. Lawyers negotiating development deals should trust their instincts. If the relationship is bad from the outset, or if the process of negotiating threatens to break the relationship, the best advice to the client may be to walk away. A significant development effort requires trust and cooperation, and no piece of paper—no matter how many "time is of the essence" clauses it contains—can mend it. It is best to put one's legal energies into structuring the payment and incentive schemes of the deal; they are most likely to drive the relationship to a successful result. In other words, each side should probably have some skin in the game in order to take the project seriously. That means hybrid payment terms are very useful in bioinformatics projects.

Do You Know Your Salesman's Birthday?—Development and Sales Cycles

The issue described in this section is not unique to bioinformatics, but it is key in any development project with a long sales cycle. Enterprise (as opposed to standard desktop) software is notorious for having a long sales cycle, and the cycle is longest for governmental, university, or quasi-governmental customers, all of which are key potential customers for bioinformatics software.

Software agreements with long sales cycles have some peculiarities. First, they tend to stop and start. Given the mobility of workers in the technology industry, a project with a contract negotiation and sales cycle of over a year may outlive the tenure of some of its salesmen and customers. That means institutional memory is lost, and issues that have been closed come open again. If you think this may happen in your deal, it is a good idea to write down some notes about how the deal got to where it is under your watch, or in any given iteration, so the reasons for particular concessions or agreements need not be lost or reconstructed later.

Second, market conditions can change within the space of a one-to-two-year sales cycle. That means suppliers may change their pricing, or customers may lose the budgets that were originally earmarked for the deal. In academic and quasi-governmental settings in particular, funding may be on a strict year-to-year cycle. Customers may actually need to spend money before the fiscal year ends, or risk losing the funding entirely.

One approach to reduce, if not solve, this concern is to develop in phases, as described above. Alternatively, a limited-term testing agreement—under which the customer can use the software or portions of it for limited purposes as the remainder of the project is completed—can be granted subject to a modest testing fee that can later be applied toward the final project fee.

GREETING THE NEW DAY

If you are familiar with software development deals but new to the bioinformatics space, it should not be difficult to adjust. All the factors described above have footprints in other software deals, but in bioinformatics deals they are more significant or pronounced. With an eye on these differences, you should be able to meet the new day. Meeting these challenges will also help you view your day-to-day deals with more sophistication and confidence.

NOTE

1. Autodesk, Inc. v. Dassault Systemes Solidworks, 685 F. Supp. 2d 1023 (N.D. Cal. 2009).

CHAPTER 5

Software Patents

Gregory J. Kirsch and Charley F. Brown

THE BASICS OF PATENT LAW AND PROTECTION

The Purpose of Patent Law

A U.S. patent is a contract between an inventor and the government. Like all contracts, both parties to a patent (the inventor and the government) agree to certain obligations, and in exchange for this agreement they expect certain benefits from the contract.

The purpose behind the patent "contract" is to encourage inventors to disclose inventions to the public. Without some form of patent protection, inventors may be reluctant or unwilling to spend time and money developing inventions, or even if they do develop inventions, they might otherwise keep the inventions secret. Because it is in society's best interest to encourage innovation and to have such innovations made public, the U.S. patent laws[1] were created to provide an incentive to inventors to innovate and to publicly disclose their inventions.

Because the ultimate purpose behind patents is public disclosure of inventions, an inventor's obligations under a patent include disclosing the invention in sufficient detail such that a person working in the particular field of technology would be able to take the disclosure and actually make and use the invention. The inventor also has an obligation to disclose the best-known way of implementing the invention.

These obligations of the inventor are quite important, because if the government is to grant patent rights to an inventor, the government needs to ensure that the inventor has upheld his or her side of the bargain. Full disclosure of the invention is therefore an important obligation of the inventor, to ensure that the public ultimately benefits from the patent.

Requirements for Patentability

To be patentable, an invention must be useful, novel (new), and nonobvious. If these tests are met, then an inventor is entitled to patent protection and the government is obligated to grant it. A patent entitles the patent owner (the inventor, the inventor's employer, etc.) to exclude others from making, using, selling, or even offering to sell the patented invention.

It is important to understand that patent rights are exclusionary in this manner—the patent owner does not necessarily have the right to make, use, sell, or offer to sell the patented invention, since the patent owner's rights might be subject to the patent rights of others. This distinction can be understood most readily with the classic stool/chair illustration. Suppose a first inventor invents and obtains a patent on a stool, and a second inventor invents and obtains a patent on a chair (which is essentially an improved stool). The first patent owner may be able to exclude the second patent owner from making, using, selling, or offering to sell both a stool and a chair, since both might be covered by the first broad patent. Thus, even though the second patent owner has a patent on a chair, he or she may be unable to make, use, sell, or offer to sell a chair without infringing the first patent. Therefore, to take advantage of his patent in the chair, the second patent owner would have to obtain permission from the first patent owner. It is for this reason that patent rights are exclusionary in nature and govern only the rights the patent owner has with respect to others.

THE PATENT APPLICATION PROCESS

Prior Art

Patent attorneys often recommend to their clients that a search for "prior art" be performed before a decision is made as to whether to apply for patent protection. Prior art is essentially those prior published documents and activities that serve as evidence of the state of the art. Since a patentable invention must be not only useful but also new and nonobvious, the U.S. Patent and Trademark Office (USPTO), the agency of the U.S. government responsible for reviewing patent applications and granting patents, will compare the patent application to the "prior art" to determine whether the latter two tests are satisfied. If the invention as claimed in the patent application is deemed to be new and nonobvious over the prior art, then the USPTO will grant a patent.

Traditionally, prior art searches have been performed by professional searchers who manually search the records of the USPTO for prior issued patents and other technical literature. This type of search may be sufficient for simple inventions, but with more complex inventions (such as biotechnology and software), and inventions in fast-moving technologies (such as for Internet-related technology), traditional searching techniques are often inadequate. Because prior art may be found in commercially available online databases (such as Lexis and Dialog) and on the Internet, as well as in specialized databases, these additional search locations also may be explored for search purposes.

Once the results of the search(es) are in hand, the patent attorney often is asked to render an opinion of patentability—that is, whether the invention is likely patentable over the known prior art, and if so, what aspects of the invention might be patentable. This opinion serves to aid the inventor in determining whether a patent application is justified, and if so, it will help the patent attorney in due course to determine how broadly to claim the invention in the patent application.

Timing of Patent Application

In the United States, the timing of the filing of a patent application can be critical. If the filing of the patent application is delayed too long, then patent protection may be barred, and the invention effectively donated to the public domain. Specifically, according to current

U.S. patent law, a patent application must be filed within one year of any one of the following events:

- The disclosure of the invention in a publication
- The public use of the invention
- The patenting of the invention by another
- An offer to sell the invention in the United States (regardless of whether a sale is ultimately made)

Therefore, if any one of these events occurs, then a one-year timer begins, and if a patent application is not filed within the one-year period, patent protection is barred. This can be problematic for companies that are trying to market or otherwise publicize their inventions, as it forces a decision to be made soon thereafter as to whether to pursue patent protection.

It is important to note that the United States currently has some of the more lenient rules regarding timing, and that most other countries have no such one-year grace period. For example, in Europe, if an invention is described in a publication or is otherwise publicized before a patent application is filed, then patent protection is barred. Thus, for companies interested in patent protection outside the United States, the lowest common denominator requires that a patent application be filed before *any* publication or public disclosure of the invention, in order to maintain the possibility of patent protection in foreign jurisdictions.

The patent laws of the United States are currently in transition, and the rules regarding timing of patent applications in light of prior activity will soon be significantly changing. In the autumn of 2011, the U.S. Congress passed into law a major overhaul of the U.S. patent system, which will go into effect over time.[2] Among the new provisions of U.S. patent law are a change from a "first-to-invent" system to a "first-inventor-to-file" system. U.S. patents have traditionally been awarded to inventors who could prove they invented first. However, under new U.S. patent law, effective as of March 16, 2013, if two inventors file their U.S. patent applications around the same time, the patent will be awarded to the inventor who filed first (with some limited exceptions), regardless of who invented first.

The Patent Application

If the inventor (or the inventor's employer) decides to move forward with a patent application, the inventor and the patent attorney begin working together to draft a patent application. A regular utility patent application contains two parts: (1) a detailed specification describing how to make and use the invention, and (2) one or more patent claims, which define the scope of the invention.

The detailed specification can be thought of as the proof that the inventor actually knows how to make and use the invention. While there is no requirement that the inventor actually build a prototype of the invention, the written specification must be sufficiently detailed such that an ordinary designer in the field of the invention (such as an ordinary software programmer, in the case of software-related inventions) would be able to review the specification and actually make and use the invention, without further input from the inventor. The specification need not be a basic tutorial in the relevant technical field; it can assume that the reader has a basic understanding of the technical field and has access to reference material as needed. However, the specification cannot leave out any substantive details of the invention, and the best-known way for implementing the invention also must be included.

The patent claims are sentences that define the scope of the invention, and they are analogous to the "metes and bounds" of a real estate land deed. In the same way that any deed for the sale of land will include a legal description of the land owned, a patent claim comprises a sentence that defines the essential elements of the invention protected by the patent. To infringe a patent, a person or entity must make, use, sell, or offer to sell a product, or use a process, that falls within the scope of at least one claim of the patent. If one or more elements of a claim of a patent are missing, then there is no infringement of that patent claim.

At least some of the patent claims are initially drafted so that they define the invention as broadly as possible, with other claims defining the invention more narrowly. Generally speaking, shorter patent claims are broader, because they have fewer "limitations" in them. To be valid, patent claims must find "support" in the specification—that is, the claimed invention must be described in detail in the specification.

Once the patent application is drafted, it is filed with the USPTO. Ultimately, a patent examiner, an employee at the USPTO with technical or scientific expertise in the field of the invention, and with training to review patent applications, is assigned to review the patent application. The examiner initially reviews the patent application and performs a search for prior art related to the claimed invention. The patent examiner thereafter compares the patent claims to the prior art and makes an initial determination as to whether the claimed invention satisfies the tests for patentability—useful, new, and nonobvious. If the invention is deemed useful, new, and nonobvious over the prior art, then the patent application is allowed to issue as a patent.

More often than not, the patent examiner will reject the patent claims as being too broad, or having some syntactical problems. A written "office action" is mailed from the USPTO to the patent attorney, outlining the rejections and inviting the patent attorney to respond. The patent attorney and the inventor may thereafter review the office action, as well as the cited prior art, and respond by either amending the patent claims and/or presenting one or more arguments that the patent claims are in fact patentable. This negotiation between the inventor (and/or the inventor's patent attorney) and the patent examiner is repeated, until (hopefully) the patent examiner allows the patent application to issue. This process is referred to as "patent prosecution."

In the United States, while the patent application is pending (before issuance as a patent), it is generally maintained in secrecy by the USPTO for at least 18 months. If the patent application is filed in foreign countries or jurisdictions, then the patent application is published 18 months after the original filing date. If the patent application is filed just in the United States, then if the inventor so chooses, the patent application will be published only after the patent actually is granted.

Issuance of the Patent

Once a patent issues (grants), it is made public and is enforceable by its owner against others. Pending patent applications may not be enforced against alleged infringers, although in certain cases damages may be recovered retroactively to the publication date of the patent application, if the claims have not materially changed between the publication date and the grant date. The patent owner may use an issued patent to extract a royalty or other compensation from an infringer, or may even attempt to force an infringer to stop infringing altogether. If the patent owner can't persuade an accused infringer to pay or to stop infringing, a lawsuit may be filed in federal court to enforce the patent rights, to recover monetary compensation, or to pursue an injunction or temporary restraining order against the infringer.

Because a patent application is reviewed and examined by a patent examiner before a patent issues, it is presumed to be valid by the courts. While patents can be invalidated for a variety of reasons, it is an accused infringer's burden to prove that a patent is invalid and should not have been granted. Some of the reasons that a patent may be invalidated include: material prior art not previously considered by the patent examiner, insufficient disclosure of the invention, or fraud perpetrated on the USPTO by the inventor or the patent attorney.

Finally, after issuance of a patent, the USPTO requires periodic maintenance fees to be paid, to keep the patent active. Specifically, within four, eight, and 12 years after issuance of a patent, escalating fees must be paid to the USPTO; otherwise, the patent lapses and becomes unenforceable.

PATENTS FOR SOFTWARE

Legal Environment

Many high-tech companies use patents for legally protecting their computer software and other innovative techniques from misappropriation by others. At the same time, many software companies have grown to rely less on copyright protection to protect their innovations. This divergent, but possibly related, trend has caused a dramatic shift in thinking in the high-tech industries—a change destined to have a major impact on the continued advancement of these technologies.

As discussed, patent law requires that an invention be novel (new) and nonobvious. To be novel and nonobvious, the invention must not have been a part of the "prior art," and it must not have been an obvious variation of the prior art, regardless of whether the invention was created independently from that of the prior art. An issued patent allows the patent owner to exclude others from making, using, or selling the patented invention. Such a right operates regardless of whether another party copied the invention from the inventor of the patented subject matter.

The allowable scope of patent protection for software-related inventions has changed considerably. Prior to the early 1980s, conventional wisdom held that software was not patentable. This view was furthered by two notable U.S. Supreme Court cases, *Gottschalk v. Benson*[3] (1972) and *Parker v. Flook*[4] (1978), in which software-implemented inventions were held not to be patentable.

Although Congress has never passed a patent statute expressly providing for federal patent protection of computer software, in 1981 the U.S. Supreme Court decided *Diamond v. Diehr*,[5] which expressly held for the first time that computer software was patentable (or put another way, that an invention was not necessarily unpatentable simply because it utilized software). In *Diehr*, the Supreme Court found that computer software was patentable, provided that any claims to the computer software were not merely a procedure for solving a mathematical formula. Though the significance of this case was not immediately appreciated by most computer software companies, certain companies recognized that the Supreme Court had provided a road map for patenting computer software. In addition to the Supreme Court's opening of the door for protecting computer software with patents, the Court of Appeals for the Federal Circuit (the Federal Circuit)[6] was established in 1982 as the sole appellate federal court authorized to hear all patent cases. The Federal Circuit promptly made it clear that patent protection for software would assume greater prominence in the future.

During the mid- to late-1980s, the preferred means of protecting computer software began to shift from copyrights to patents. Specifically, more companies began to file patent applications for their computer software inventions as the USPTO began to relax its standards somewhat for issuing computer software patents and the federal courts upheld the majority of patents issued.

However, it was not until the early 1990s that many software developers began favoring patents for computer software. The reason for the change stemmed in part from a series of federal court decisions, led by *Computer Associates v. Altai*[7] as well as *Lotus v. Borland*,[8] which ruled that copyright protection for computer software should be applied narrowly to essentially protect little more than the exact copying of software code. These courts sent the message that protection for the functionality of computer software should be sought under patent laws.

The trend toward broader patents directed to software inventions was furthered in cases such as *In re Alappat*,[9] decided by the Court of Appeals for the Federal Circuit in 1994. In *Alappat*, the Federal Circuit judges decided that the subject patent application presented proper "useful" statutory subject matter, even though the patent claims merely presented the software process in terms of different physical "elements" within a machine to perform the functions. In *In re Lowry*,[10] the Federal Circuit held that a claim to a data structure stored on a computer-readable medium that increases computer efficiency was statutory subject matter. However, in *In re Warmerdam*,[11] a claim to a data structure was held per se nonstatutory subject matter where the data structure did not cause a functional change in the computer.

In 1998, in *State Street Bank v. Signature Financial Group*,[12] the Federal Circuit ruled that a patent directed to a "hub and spoke" data-processing system that allows mutual funds (spokes) to pool their assets into an investment portfolio (hub) organized as a partnership was valid. The claimed invention made several calculations on a daily basis related to the financial configuration of the mutual funds, including each fund's percentage share of the investment's assets and expenses. In ruling that the subject patent was valid, the court held that a long-standing doctrine, known as the "business method exception," was ill-conceived and not appropriate. The business method exception to patentability has traditionally been used to strike down patents directed to methods of doing business. The court ruled that this exception to patentability is not only ill-conceived, but it is not appropriate to computer-implemented inventions that clearly otherwise satisfy the tests of patentability.

In the later part of the first decade of the 2000s, the Federal Circuit's view regarding software and business method patents began to narrow somewhat. In 2007, in *In re Comiskey*,[13] the Federal Circuit decided that claims relating to mandatory arbitration in certain legal contracts were unpatentable subject matter under Section 101. The claims at issue related entirely to mental processes (i.e., used human intelligence) and, as such, were unpatentable, while claims to the same process but requiring a physical device or module were statutory subject matter. However, the court noted that "[t]he routine addition of modern electronics to an otherwise unpatentable invention typically creates a *prima facie* case of obviousness."[14]

On October 30, 2008, the Federal Circuit decided *In re Bilski*.[15] The patent application at issue was titled "Energy Risk Management Method," and had method claims directed to a process for a broker to hedge risks in commodity transactions. The majority decision established a single test for determining patentability of a process. The test, known as the "machine-or-transformation test," states that a process is patentable if:

1. it is tied to a particular machine or apparatus, or
2. it transforms a particular article into a different state or thing.[16]

Applying the machine-or-transformation test to Bilski's claims, the Federal Circuit found that since the claims failed to recite a machine, it was unnecessary to decide any issues relating to the machine-tying prong of the test. The Federal Circuit stated, "[w]e leave to future cases the elaboration of the precise contours of machine implementation, as well as the answers to particular questions, such as whether or when recitation of a computer suffices to tie a process claim to a particular machine."[17] Next, the Federal Circuit turned to the transformation prong of the test, and found the claims unpatentable as they did not "transform any article to a different state or thing."[18] The Federal Circuit made clear that transformation of "legal obligations or relationships, business risks, or other such abstractions" will fail the machine-or-transformation test because they are neither physical objects or substances nor representative of the same.[19] The ruling narrowed software and business method patentability somewhat, in situations where process claims recited steps not necessarily performed by a computer.

In June of 2010, the U.S. Supreme Court rejected the Federal Circuit's holding in *In re Bilski* that the machine-or-transformation test is the *sole* test to determine whether a particular process constitutes patent-eligible subject matter.[20] The Supreme Court did, however, view the test as a helpful indicator in patentability analysis. Like the Federal Circuit, the Supreme Court majority held the Bilski claims to be unpatentable subject matter as they were directed to an abstract idea, referring back to the Supreme Court decisions in *Gottschalk v. Benson* and *Parker v. Flook*. In so doing, the Supreme Court rejected the patentable subject matter analysis set forth in *State Street Bank*, although it held that the machine-or-transformation test may be used as a nonexhaustive tool in determining patent eligibility of processes.

The definition of patentable subject matter continues to develop post-*Bilski*. In *CyberSource*,[21] the following method claim for detecting fraud in credit card transactions over the Internet by comparing IP addresses was found to be patent-ineligible:

> 3. A method for verifying the validity of a credit card transaction over the Internet comprising the steps of:
>   a) obtaining information about other transactions that have utilized an Internet address that is identified with the [ ] credit card transaction;
>   b) constructing a map of credit card numbers based upon the other transactions and;
>   c) utilizing the map of credit card numbers to determine if the credit card transaction is valid.[22]

The claim does not require a specific algorithm and does not require any physical device. The patent also contained a "Beauregard claim," named after the inventor in *In re Beauregard*.[23] A Beauregard claim is a claim to a computer-readable medium (e.g., a disk, hard drive, or other data storage device) containing program instructions for a computer to perform a particular process. The Beauregard claim in *Cybersource* repeated the limitations of the method claim but embodied on a computer-readable medium.

The Federal Circuit applied the machine-or-transformation test and found the method claim lacking. The court stated that "collection and organization of data regarding credit card numbers and Internet addresses is insufficient to meet the transformation prong of the test."[24] Turning to the machine prong of the test, the patent applicant argued that the method was tied to a particular machine because it "would not be necessary or possible without the Internet."[25] The Federal Circuit, however, found that the Internet cannot perform the fraud detection steps of the claimed method and that nothing in the method claim requires an infringer to use the Internet to obtain data (as opposed to obtaining the data from a pre-compiled database). The court viewed the Internet as the data source and repeated its holding from *In re Grams*, "[data-gathering] step[s] cannot make an otherwise nonstatutory claim statutory."[26]

Having determined that the method claim failed to meet the machine-or-transformation test, the Federal Circuit heeded the Supreme Court ruling in *Bilski* and went beyond the machine-or-transformation test, and determined that the method claimed was not patent-eligible subject matter as it was drawn to "an unpatentable mental process—a subcategory of unpatentable abstract ideas."[27] The Federal Circuit cited to *Gottschalk* and *Flook*, and held that because the method claim sought "to patent the use of human intelligence in and of itself," the claims were drawn to abstract ideas and were invalid.[28] The Federal Circuit explained that all of the method steps can be performed in the human mind, or by a human using a pen and paper.[29]

The Federal Circuit then addressed the *Beauregard* version of the method claim and found it to be patent-ineligible as well. The court stated, "[r]egardless of what statutory category ('process, machine, manufacture, or composition of matter,' 35 U.S.C. § 101) a claim's language is crafted to literally invoke, we look to the underlying invention for patent-eligibility purposes." Since the underlying method was patent-ineligible as directed to a mental process, embodying that mental process on a computer-readable medium did not make an otherwise unpatentable process patentable. The court recognized its prior holdings in *In re Abele*[30] (finding a claim patentable that included both mental steps and physical steps), and *In re Alappat*[31] (programming a general purpose computer to perform an algorithm "creates a new machine, because a general purpose computer in effect becomes a special purpose computer once it is programmed to perform particular functions pursuant to instructions from program soft-ware."). However, the court recognized that a mental process does not become patent-eligible through its execution on a computer. To be patent-eligible, the machine "must impose meaningful limits on the claim's scope," so the "the incidental use of a computer to perform the mental process of claim 3 does not impose a sufficiently meaningful limit on the claim's scope."[32] Thus, claim 2 (like claim 3) was not patent-eligible under § 101.

On March 2, 2012, the Federal Circuit, in *MySpace, Inc. v. GraphOn Corp.*,[33] held that courts should avoid resolving Section 101 disputes by instead resolving disputes under the well-defined doctrines of Sections 102 (novelty), 103 (nonobviousness), and 112 (written description). The Federal Circuit upheld the district court ruling (under Sections 102 and 103) invalidating GraphOn's patents directed to creating, modifying, and searching for database records over a computer network.[34] The Federal Circuit went on to criticize the status of Section 101 jurisprudence and to discourage courts from attempting to resolve disputes under Section 101. Rather than entering the "murky morass that is § 101 jurisprudence," courts should "avoid the swamp of verbiage that is § 101 by exercising their inherent power to control the processes of litigation."[35] Avoiding Section 101 analysis would "conserve judicial resources, and bring a degree of certainty to the interests to both patentees and their competitors in the marketplace."[36]

The Federal Circuit stated that "[t]he problem with addressing § 101 initially every time it is presented as a defense is that the answer in each case requires the search for a universal truth."[37] The majority identified that "[w]hen it comes to explaining what is to be understood by 'abstract ideas' in terms that are something less than abstract, courts have been less successful."[38] Accordingly, the court held that district courts should exercise their inherent power to control the processes of litigation, "and insist that litigants initially address patent invalidity issues in terms of the conditions of patentability defenses as the statute provides, specifically §§ 102, 103, and 112."[39] The result is more efficient litigation, conservation of judicial resources, and bringing about certainty to the interests of patentees and their competitors.[40]

BIOINFORMATICS PATENTS

Bioinformatics is the science of using information to understand biology through the use of computational tools and databases. In December 1999, the USPTO created a special group of patent examiners to review bioinformatics patent applications—Art Unit 1631. Bioinformatics inventions are typically directed to computer-implemented methods for collecting, storing, processing, or analyzing biological data. As such, laws that affect software patents will affect bioinformatics patents.

What Are Bioinformatics Inventions?

Specific examples of bioinformatics inventions include:

- protein modeling systems;
- sequence alignment methods;
- combinatorial library systems;
- computer-aided sequence visualization and analysis systems;
- compositions of matter on microchips; and
- methods of comparing patients' conditions.

Other examples include:

- tools for automation of laboratory experimentation and data generation;
- the design, implementation, management, and integration of databases;
- the design of communications standards between databases from multiple sources; and
- data-mining tools for retrieving, processing, analyzing, and simulating biological data to determine a gene location within a chromosome, find similar genes or proteins from other species, and determine 3D structure and function of different proteins.

Problems in Obtaining Bioinformatics Patents

Most bioinformatics inventions are software processes and therefore rely on method, system, or computer-readable medium claims. However, some bioinformatics inventions can utilize composition of matter or article of manufacture claims when the value of the invention resides in a therapeutic product, rather than software used to identify the product. Of course, in the event a patentable therapeutic product is invented using bioinformatics tools, a patent application may also be filed on the resulting product. Through creative use of so-called product-by-process claims (defined in further detail later), protection may be extended to products created as a result of the bioinformatics tools.

As in other industries, an effective patent claim strategy should be devised to account for the various business models used in the industry. For example, some companies are engaged in the licensing of software, access to databases, hardware, and diagnostic tests. In this case, patent claims should be drafted to capture all possible business uses of the invention to have a greater impact on the market, and they should also be drafted so as to require a sole infringer to simplify enforcement.

Bioinformatics is an interdisciplinary science. Even though the USPTO established a bioinformatics art unit, not all bioinformatics inventions may find their way to the USPTO patent examiners in this examining unit. It can be challenging for the USPTO to identify and hire patent examiners that are trained in both computer science and in biology, and the USPTO therefore in some cases relies on examiners in other groups to review bioinformatics-related patent applications.

For example, a bioinformatics patent application recently examined by the USPTO contained claims directed to methods for managing a biological database. In short, the claimed invention probed multiple databases using a set of biological sequences and collected various types of data related to the sequences. These data included functional data, structural data, immunohistochemical data, in situ hybridization data, and expression data. Among its various teachings, the patent application explained (1) that functional data was used to assign a biological sequence into one of 12 putative functional categories (e.g., metabolism, signaling, etc.), (2) that structural data related to three-dimensional macromolecular structures (e.g., proteins and polynucleotides), (3) that immunohistochemical data included information on antibodies as well as confocal images of the immunohistological patterns obtained with the various antibodies, and (4) that in situ hybridization data provided tissue- or cell-specific expression patterns of genes.

Throughout examination, the examiner at the USPTO demonstrated his lack of proficiency with respect to the biological arts. For example, the examiner erroneously equated structural data relating to the three-dimensional structure of polypeptides and polynucleotides with structures inside a body such as a muscle, a kidney, or a lung. Furthermore, as the examiner was unfamiliar with genes and amino acids, the examiner did not understand the term expression data. At one point, the examiner stated that immunohistochemical data represented the record of an individual's immunization. Applicants continuously argued that terms such as immunohistochemical, in situ, and expression were terms of the biological arts that the skilled person would understand without hesitation.

During a phone interview between the patent attorney and the examiner, the examiner stated that he had tried on several occasions to have the case reassigned to an examiner more familiar with the biological subject matter. Unable to execute the transfer, and finding no prior art that would prevent issuance of a patent, the examiner allowed the case but demanded that these ordinary terms of art be defined in each independent claim to include the term putative. The examiner stated that both the examiner and the supervisory primary examiner were unfamiliar with that term and, therefore, it had to be defined in the claim. The resulting claims contain definitions that would have been wholly unnecessary had the examiner been familiar with both computer and biological sciences.

Useful Claim Types for Bioinformatics Inventions

Methods/Computer-Readable Media

The majority of bioinformatics patents will have method claims. While the claims will have to pass the novelty and nonobviousness requirements described earlier in this chapter, the most challenging aspect to claim drafting relates to patentable subject matter.

In *Ex parte Kelkar,*[41] the USPTO's Board of Patent Appeals and Interferences (BPAI) ruled on whether certain bioinformatics method and computer-readable medium claims defined patent-eligible subject matter. The claims were directed to a method of determining similarity between two genetic profiles and a computer program stored on a medium for performing the method.

> 1. A method for determining similarity between portions of gene expression profiles in a computer comprising the steps of:
>
> processing a number of gene expression profiles with a similar sequences algorithm that is a time and intensity invariant correlation function to obtain a data set of gene expression profile pairs and a match fraction for each gene expression profile pair;
>
> listing gene expression profile pairs in clusters by their match fractions;
>
> removing a first gene expression profile from a cluster when another cluster has another gene expression profile with a higher match fraction with the first gene expression profile, unless the other gene expression profile requires a larger number of subsequences to achieve similarity with the first gene expression profile;
>
> repeating the removing step until all gene expression profiles are listed in only one cluster;
>
> providing output of the listing of clusters of gene expression profiles.[42]

The BPAI held that the claims did not satisfy the machine-or-transformation test because the method failed to require any particular machinery and failed to transform an article into a different state or thing. The BPAI cited *Flook* and held that the claim was improperly directed to an abstract algorithm because the underlying innovation was an unpatentable mathematical algorithm. Despite the fact that the preamble states that the method is performed "in a computer," the BPAI found that limitation to be "a field-of-use limitation that is insufficient to render the otherwise ineligible process . . . patent eligible."[43]

The BPAI also held that the computer-readable medium claims were not directed to patent-eligible subject matter. Citing *Nuijten*, the BPAI held that the computer-readable medium limitation encompassed unpatentable "carrier wave storage."[44] Specifically, the BPAI stated:

> When the broadest reasonable interpretation of a claim covers a signal per se, the claim must be rejected under 35 U.S.C. § 101 as covering nonstatutory subject matter. See *In re Nuijten*, 500 F.3d 1346, 1356–57 (Fed. Cir. 2007) (transitory embodiments are not directed to statutory subject matter). See also, Subject Matter Eligibility of Computer Readable Media, 1351 Off. Gaz. Pat. Office 212 (Feb. 23, 2010).[45]

These cases underscore the importance of properly defining the invention in the claims that accompany a patent application. Case law provides specific requirements that must be met for a claimed invention to be deemed statutory subject matter. The "machine-or-transformation" test created in *Bilski*, as well as the prohibition in *Nuijten* against claim coverage of propagated signals per se, provide relatively bright-line tests for ensuring that a claimed invention comprises potentially patentable subject matter.

Databases

The Federal Circuit and the USPTO have declared that a database is merely a "data structure" or "nonfunctional descriptive material," and is therefore not patentable.[46] However, even if the database itself does not constitute patentable subject matter, the manner of creating and managing the database may constitute a patentable process.[47]

Product by Process

A product-by-process claim defines a product through the use of a particular method used to create the product. A true product-by-process claim is focused on the product.[48] Product-by-process claims will be invalid if the resultant product is known in the art. However, a product-by-process claim will be infringed only if the infringer is making the claimed product according to the method. The Federal Circuit explains in *Amgen Inc. v. F. Hoffmann-La Roche, Ltd.*:

> In determining validity of a product-by-process claim, the focus is on the product and not on the process of making it. See Atl. Thermoplastics Co. v. Faytex Corp., 970 F.2d 834, 841 (Fed. Cir. 1992) (explaining that, in BASF, the validity rule "focused on the product with less regard for the process limits"); Brown, 459 F.2d at 535 (focusing on the product claimed and not the process); Pilkington, 411 F.2d at 1348 (noting that the product itself must be new). That is because of the already described, long-standing rule that an old product is not patentable even if it is made by a new process. As a result, a product-by-process claim can be anticipated by a prior art product that does not adhere to the claim's process limitation. In determining infringement of a product-by-process claim, however, the focus is on the process of making the product as much as it is on the product itself. See Abbott Labs. v. Sandoz, Inc., 566 F.3d 1282, 1293 (Fed. Cir. 2009) (en banc). In other words, "process terms in product-by-process claims serve as limitations in determining infringement." Id. (quotation marks omitted). As a result, a product-by-process claim is not infringed by a product made by a process other than the one recited in the claim. Id.[49]

Due to the confusion surrounding product-by-process claims, they should not be relied upon as the sole means for protecting novel products. However, a savvy patent applicant, armed with a sufficiently enabling disclosure, can make much use of product-by-process claims.

CONCLUSION

The legal landscape of software patents, and thereby bioinformatics patents, is constantly evolving. Similarly, the technological underpinnings of bioinformatics are moving forward at a rapid pace. Thus, patenting new bioinformatics inventions requires overcoming constant challenges ranging from the patentability of the inventions to the interdisciplinary technological background required of both patent prosecutors and the PTO. Patent prosecutors will need to stay up to date with the current law and technology, and be agile in claiming strategies, to ensure adequate protection for new bioinformatics inventions.

EXAMPLES OF BIOINFORMATICS PATENTS

Patent Number: U.S. Pat. No. 5,557,535

Title: "Method and system for protein modeling"

Abstract: A method in a computer system for modeling a three-dimensional structure of a model protein is provided. In a preferred embodiment, the modeling is based upon a three-dimensional structure of a template protein and an amino acid sequence alignment of the model protein and the template protein. The proteins comprise a plurality of amino acids having backbone atoms and side chain atoms. For each amino acid in the model protein, when the template protein has an amino acid aligned with the amino acid of the model protein, the position of each backbone atom of the amino acid of the model protein is established based on the position of a topologically equivalent backbone atom in the aligned amino acid of the template protein. The inter-atomic distance constraints for each pair of atoms with an established position is generated. Finally, the position of each atom in the model protein is set so that the inter-atomic distances are in accordance with the constraints.

Claim 1: 1. A method in a computer system for modeling a three-dimensional structure of a variable region of a model protein, the model protein having amino acids, the amino acids having positions within a three-dimensional structure, the method comprising the steps of:

receiving relative positional information between pairs of amino acids;

establishing a position for a first amino acid of the variable region; and

for each amino acid pair in the variable region, generating a model position for the amino acids based on the received relative positional information for the pair of amino acids.

Patent Number: U.S. Pat. No. 6,023,659

Title: "Database system employing protein function hierarchies for viewing biomolecular sequence data"

Abstract: Disclosed is a relational database system for storing biomolecular sequence information in a manner that allows sequences to be catalogued and searched according to one or more protein function hierarchies. The hierarchies allow searches for sequences based upon a protein's biological function or molecular function. Also disclosed is a mechanism for automatically grouping new sequences into protein function hierarchies. This mechanism uses descriptive information obtained from "external hits," which are matches of stored sequences against gene sequences stored in an external database such as GenBank. The descriptive information provided with the external database is evaluated according to a specific algorithm and used to automatically group the external hits (or the sequences associated with the hits) in the categories. Ultimately, the biomolecular sequences stored in databases of this invention are provided with both descriptive information from the external hit and category information from a relevant hierarchy or hierarchies.

Claim 1: 1. A computer system comprising:

a database containing records pertaining to a plurality of biomolecular sequences;

a first hierarchy of protein function cate gories into which at least some of said biomolecular sequences are grouped, said protein function categories specifying biological functions of proteins corresponding to said biomolecular sequences and said first hierarchy including

(i) a first set of protein function categories specifying biological functions at a cellular level, and
(ii) a second set of protein function categories specifying biological functions at a level above the cellular level; and

a user interface allowing a user to selectively view information regarding said plurality of said biomolecular sequences as it relates to said first hierarchy.

Patent Number: U.S. Pat. No. 6,094,626

Title: "Method and system for identification of genetic information from a polynucleotide sequence"

Abstract: A method and system for the identification of genetic information from a polynucleotide sequence is described. Genetic information from a raw polynucleotide sequence is identified by assigning values to nucleotide base changes along the raw polynucleotide sequence and processing these values. Particularly, a distribution of the values of the base changes is generated and the variance from a random distribution of this distribution is calculated by determining the root mean square variance of the distribution. The types of genetic information that can be identified using this system and method include the presence of a protein-encoding region, or gene; the number of genes in a given overall polynucleotide sequence; the reading frame for the gene sequence; and tRNA and rRNA sites.

Claim 1: 1. A method of identifying genetic coding regions in an isolated polynucleotide sequence comprising the steps of:
(a) entering the polynucleotide sequence into a data processing system;
(b) identifying base changes in the polynucleotide sequence and assigning an observed base change value to each base change;
(c) transferring the observed base change values to a rate of change matrix;
(d) generating a distribution of the observed base change values from the rate of change matrix;
(e) comparing the distribution of the observed base change values to expected base change values in a statistically random biologically weighted distribution to determine one or more variances; and
(f) displaying the variances, which correspond to the genetic coding regions, to identify those genetic coding regions.

Patent Number: U.S. Pat. No. 6,518,487

Title: "Cyclin D polynucleotides, polypeptides and uses thereof"

Abstract: The invention provides isolated polynucleotides and their encoded proteins that are involved in cell cycle regulation. The invention further provides recombinant expression cassettes, host cells, transgenic plants, and antibody compositions. The present invention provides methods and compositions relating to altering cell cycle protein content and/or composition of plants.

Claim 1: 1. An isolated nucleic acid encoding a polypeptide that increases transformation efficiency, comprising:
(a) a polynucleotide that encodes a polypeptide SEQ ID NO: 2;
(b) a polynucleotide having at least 95 percent identity to the entire coding region of SEQ ID NO: 1 wherein the percent identity is determined by GCG/bestfit program using a gap creation penalty of 50 and a gap extension penalty of 3;

(c) a polynucleotide characterized by the sequence set forth in SEQ ID NO: 1;
(d) a polynucleotide complementary to a polynucleotide of (a) through (c); or
(e) a polynucleotide having the sequence of ATCC deposit having the Designation No. 98848.

NOTES

1. Title 35, United States Code.
2. Leahy-Smith America Invents Act (AIA), Public Law 112–29, 125 Stat. 284 (Sept. 16, 2011); Technical Corrections to the AIA, Public Law 112–274 (Jan. 14, 2013).
3. 409 U.S. 63 (1972).
4. 437 U.S. 584 (1978).
5. 450 U.S. 175 (1981).
6. United States Court of Appeals for the Federal Circuit, http://www.cafc.uscourts.gov/.
7. 982 F.2d 693 (2d Cir. 1992).
8. 516 U.S. 233 (1996).
9. 33 F.3d 1526 (Fed. Cir. 1994).
10. 32 F.3d 1579 (Fed. Cir. 1994).
11. 33 F.3d 1354 (Fed. Cir. 1994).
12. 149 F.3d 1368 (Fed. Cir. 1998).
13. 499 F.3d 1365 (Fed. Cir. 2007).
14. *Id.* at 1380. *See also In re* Nuijten 500 F.3d 1346, 1356–57 (Fed Cir. 2007) (transitory embodiments are not directed to statutory subject matter).
15. 545 F.3d 943 (Fed. Cir. 2008).
16. *Id.* at 961.
17. *Id.* at 962.
18. *Id.* at 963.
19. *Id.*
20. Bilski v. Kappos, 130 S. Ct. 3218 (2010).
21. 654 F.3d 1366 (Fed. Cir. 2011).
22. *Id.*
23. 53 F.3d 1583 (Fed. Cir. 1995).
24. Cybersource Corp. v. Retail Decisions, Inc., 654 F.3d 1366, 1370 (Fed. Cir. 2011).
25. *Id.*
26. *In re* Grams, 888 F.2d 835, 840 (Fed. Cir. 1989).
27. *Cybersource*, 654 F.3d at 1372.
28. *Id.* at 1371.
29. *Id.*
30. 684 F.2d 902, 908 (C.C.P.A. 1982).
31. 33 F.3d 1526, 1545 (Fed. Cir. 1994).
32. *Cybersource*, 654 F.3d at 1375.
33. 672 F.3d 1250 (Fed. Cir. 2012).
34. *Id.* at 12.
35. *Id.* at 17.
36. *Id.*
37. *Id.* at 13.
38. *Id.* at 14.
39. *Id.* at 1260.
40. *Id.* at 17.
41. App. No. 2009-004635 (B.P.A.I., September 24, 2010).
42. U.S. Pat. App. No. 10/629,448, Pub. No. 20091231234 (published Feb. 3, 2005).
43. U.S. Pat. App. No. 2009-004635 at p.4 (B.P.A.I., Sept. 24, 2010).
44. *Id.* at 5.
45. *Id.* at 4.
46. *In re* Warmerdam, 33 F.3d 1354 (Fed. Cir. 1994); *In re* Lowry, 32 F.2d 1579 (Fed. Cir. 1994).
47. *See, e.g.*, U.S. Patent No. 6,023,659.
48. *See, e.g.*, U.S. Patent No. 6,518,487.
49. 580 F.3d 1340, 1369–70 (Fed. Cir. 2009).

CHAPTER 6

A Gene Patent Primer

Jennifer Carter-Johnson

INTRODUCTION

Patent Overview

Patents are time-limited exclusive rights granted to inventors in order to encourage advancement of the technology.[1] While there are three types of patents, utility patents are by far the most common type of patent and are the patents most likely to apply to gene-related patents;[2] therefore, this chapter will discuss only utility patents. Chapter 35 of the United States Code (the Patent Act) enumerates the rights of the patent owner and the requirements for obtaining patent protection.[3] A patent gives the owner the right to prevent others from making, using, selling, or offering to sell the owner's patented invention within the United States or importing the invention into the United States.[4] Inventions must be new, useful, and nonobvious in order to qualify for patent protection.

Patents are an important societal pact. The inventor receives a patent in exchange for disclosing his or her invention, which must be novel. Society banks this new information and, at the end of the patent term, is free to use the information as it will. Therefore, patent applications must not only disclose the invention but also specifically describe the invention in such a way as to enable a person skilled in the art, i.e., a person who has the normal skills and training found in the invention's technological area, to duplicate that invention at the end of the patent term. On the other hand, the inventor may use his or her monopoly to sell the invention without competition or to license others to sell the invention, i.e., to force those who would like to sell the invention to pay him or her for that right.

A patent for an invention is a property right granted to an inventor. In order to receive a patent, an inventor files an application with the U.S. Patent and Trademark Office (USPTO). During the process of applying for a patent, known as patent prosecution, a patent examiner makes sure that the invention is patentable and that the application complies with all statutory requirements. The inventor is given the chance to amend his or her application to address problems during patent prosecution, resulting in a multiyear process.[5] In spite of these opportunities to amend, less than half of patent applications mature into a patent.[6] Once the patent is granted, patent rights run for 20 years from the date of filing.[7]

A patent is divided into two main sections: the specification, which is often called the written description, and the claims. The specification discloses the invention in detail and teaches the reader how to make and use the invention. The claims are generally considered the most important part of the patent as they specifically delineate the scope of the property right, i.e., the protection granted by the patent. Like a deed for a piece of land, the claims describe the "metes and bounds" of the property, and it is the claims to which the courts primarily look when determining whether a patent has been infringed.

Patent rights are negative rights; that is, the inventor has the right to exclude others from using his or her invention. However, the inventor may not have the positive right to practice the invention himself or herself. For example, imagine the invention of a pencil by Nick. Nick, the inventor of a graphite and clay rod wrapped in wood that can be used to write (the pencil), could get a patent that would give him the right to stop others from making, using, or selling pencils without his permission. Since there are no other patents in our imaginary space, Nick would be able to make pencils and sell them to others. Now imagine that Mary, seeing the pencil but wishing for error correction, invents the eraser. She also gets a patent and can produce erasers and stop others from producing the eraser. Mary would also be able to make and sell her eraser because Nick's pencil patent is unrelated to her eraser invention even though the two may be used together.

Improvements on existing patented inventions are allowed and are patentable. Frank notices that it would be more efficient to place the eraser on the nonwriting end of the pencil. Frank applies for and receives a patent for his invention.[8] But Frank cannot make his pencil/eraser combination himself without permission from Nick and Mary because making Frank's invention would require making both the pencil and the eraser. Therefore, Frank's invention infringes both Nick's and Mary's patents. Using his patent rights, Frank can, however, prevent anyone else from making his invention of a pencil with an eraser affixed to it, even Nick and Mary. Frank's patent is called a blocking patent.

Large companies often develop a patent arsenal not only to take advantage of the monopoly patents provide by selling products but also to build a defensive thicket of patents around the technology to be sold. A potential result of the large number of patents in a technology space is the potential for an anti-commons effect. The anti-commons effect occurs when the patent rights to a class of inventions is split among multiple patent holders so that no one can actually develop and sell a product without infringing someone else's patent. For example, if Company A is accused of infringing the patent of Company B, Company A likely owns a patent that Company B is infringing. In a best-case scenario, assertion of such blocking patents against Company B results in a cross-licensing agreement between the two companies rather than patent liability for Company A. However, in an anti-commons situation, the patents are split between too many companies and cross-licensing negotiations fail. The result of the anti-commons is that no products come to market and research halts until the patents expire.

Gene-Related Patents in Bioinformatics

Gene-related patents become quite important in the bioinformatics sphere. As described in Chapter 1, genes and genetic information are important inputs into most bioinformatics projects. If the project uses patented genes or genetic methods, then the researchers are likely infringing the patents and may be liable for damages or be forced to stop work. Additionally, as much of the bioinformatics research is directed at finding genes causing diseases or therapeutic targets, the patentability of genes and uses of genes may be the basis of many products of bioinformatics.

The societal exchange underlying patents is in place in order to give incentives for a person both to invest time and money into an invention and then to patent that invention rather than keep it secret. A fundamental question in patent law is how to balance those monopoly incentives with the need for information to be available for future innovators to conduct research. In fast-moving technological spaces, locking a technology into a monopoly for 20 years could give one company control over the direction of an entire industry. In the other extreme, splitting the patents to a technology into too many hands could result in an anti-commons situation and freeze an industry or new technology for many years.

There is much debate about where to draw this patent line in the realm of gene-related patents. Patents directly claiming genes isolated from humans come under the most fire. According to some research, as much as 20 percent of known genes have been patented. Some have argued that allowing genes to be patented locks away too much technology that could be used as basic research tools and creates a patent thicket that could lead to an anti-commons effect.[9] Others believe that gene patents should be outlawed for ethical, social, or moral reasons. Still others argue that patent rights have not hindered research or access to medicine and are the incentives that drive a thriving industry.[10]

PATENTABLE SUBJECT MATTER

Classes of Patentable Subject Matter

Section 101 of the Patent Act defines the classes of inventions for which patents may be granted. A patent may be filed for a "new and useful process, machine, manufacture, or composition of matter, or any new and useful improvement thereof."[11] The Supreme Court has consistently interpreted this statutory language broadly, upholding the idea that patent laws should be given wide scope.[12] In *Diamond v. Chakrabarty*, the Supreme Court held that patentable subject matter includes "anything under the sun made by man."[13]

Although the courts have read the Patent Act to include such a broad category of patentable subject matter, there are limitations. Three judicially created exceptions exist to Section 101's broad patent-eligibility principles. Laws or products of nature, physical phenomena, and abstract ideas are not patentable.[14] While these exceptions may not be required by the statutory text, the Court views the concepts covered by these exceptions as "part of the storehouse of knowledge of all men . . . free to all men and reserved exclusively to none."[15] Instead, to be patentable, an invention must be an application of a law of nature or abstract idea.[16] Unfortunately, the line between a law of nature and its application is often very hard to distinguish.

Recently, the Supreme Court has taken an interest in the Federal Circuit's patentable subject matter jurisprudence, offering some guidance in the case of *Mayo Collaborative Services v. Prometheus Laboratories, Inc.*[17] Prometheus was the exclusive licensee of patents directed at adjusting drug dosages for personalized medicine.[18] The patent claimed a method of adjusting a drug dosage by measuring the metabolites of the drug in the patient's blood and altering dosage depending on the level of metabolites. Prometheus had performed research and determined the levels of metabolite production indicative of under- and overmedication by the drug. In its patent, Prometheus claimed the steps of administering the drug to a patient and measuring the resulting metabolite levels in order to determine whether dosage of the drug needed to be altered. The Supreme Court found that the relationship between the drug and the metabolites was a newly discovered law of nature—ingesting a set amount of the drug naturally resulted in a set amount of metabolite. Therefore, the Court held that Prometheus was effectively attempting to gain patent rights to this law of nature and the patent was invalid.[19]

The Supreme Court's opinion in *Prometheus* indicates that even a newly discovered law of nature is not proper subject matter for patent protection. Importantly, the Court also held that adding steps that are "well-understood, routine, conventional activity previously engaged in by scientists in the field"[20] to that law of nature "is normally not sufficient to transform an unpatentable law of nature into a patent-eligible application of such a law."[21] This holding may have profound effects on the ability to patent isolated genes and some uses of those genes.

Types of Gene-Related Patents

Claims directed at gene-related inventions generally fall into one of two categories of patentable subject matter: a composition of matter or a method for using the gene. A claim of a gene itself, usually described by its nucleic acid sequence, would be a composition of matter claim, while a method claim would be a claim for a use of a gene, such as a method of diagnosing a disease based on a genetic mutation. While some scholars view the nucleic acid sequence of a gene as information that makes up the genetic code, current USPTO rules allow gene patenting as a composition of matter, treating the nucleic acid sequence of a gene the same as any other chemical compound.[22] Claims for uses of genes, such as for diagnostics or in screening new therapeutic drugs, are treated as any other method claim in a patent.

At the time of this writing, a case involving method claims and composition claims to human gene sequences is making its way through the legal maze. *Association for Molecular Pathology v. United States Patent and Trademark Office* revolves around the patents for two genes, BRCA1 and BRCA2, mutations of which were found to be linked to increased risk of breast and ovarian cancer. A series of patents covering these discoveries was obtained by Myriad Genetics, Inc.[23] Included in these patents were composition claims directed at the BRCA1 and BRCA2 genes themselves as well as some mutations of these genes and fragments of these genes.[24] For the most part, the claims were to *isolated* genes, rather than to the genes in their native environment of the human genome.[25] The plaintiffs brought suit, alleging that these claims were directed at patent-ineligible subject matter.

The lower court ruled that these composition claims directed at isolated human genes were indeed patent-ineligible as they were a product of nature, one of the judicially created exceptions to the broad statutory language defining what is actually patentable.[26] On appeal, the Court of Appeals for the Federal Circuit disagreed with this finding.[27] The Federal Circuit held that the claims directed at the BRCA1 and BRCA2 genes were patent-eligible because the claims were to isolated genes, meaning that the genes were no longer what was found in nature but had been altered by human intervention. Natively, the genes exist within human chromosomes in a complicated structure known as chromatin. The Federal Circuit reasoned that the act of isolation, by sequencing and reproduction of the gene, resulted in a markedly different chemical than the original, though it still contains the same information, coding for the same protein as found in the native state BRCA1 and BRCA2 genes. On further appeal to the Supreme Court, the Supreme Court vacated the Federal Circuit's decision and remanded the case back to the Supreme Court for rehearing in light of the Supreme Court's recent decision in *Prometheus*. Upon rehearing, the Federal Circuit essentially restated its prior opinion, leading to another appeal to the Supreme Court.[28]

While at the time of this writing we can't say how the Federal Circuit may hold in the rehearing, it is possible that, in light of the Supreme Court's reasoning in *Prometheus*, it may find that Myriad's claims to the isolated BRCA1 and BRCA2 genes were simply products of nature and the isolation of the genes simply a "well-understood, routine, conventional activity previously engaged in by scientists in the field."[29] If so, this would have profound impacts on the ability to patent genes, essentially closing the door to the patenting of merely isolated genes, and perhaps expanding even to those genes inserted into common vectors or other common biotechnology tools.

The Myriad patents also included method claims that were contested by the plaintiffs. Most of these method claims were directed at the process of comparing a patient's BRCA1 or BRCA2 sequence with that of the "wild-type" gene or screening potential therapeutics by comparing growth rate changes of transformed cells.[30] The vacated Federal Circuit opinion had held that the method claims directed at comparisons were not patentable subject matter due to being simply abstract ideas (another of the judicially created exceptions to patentable subject matter) while the method claims directed at therapeutic screening were patentable subject matter due to man-made nature of the transformed cells.[31] Upon rehearing, the Federal Circuit held that, in light of the Supreme Court's opinion in *Prometheus*, the gene comparison claims again were found drawn to unpatentable subject matter, this time due to claiming a law of nature, i.e., the relationship between BRCA gene mutations and risk of breast and ovarian cancer.[32] Similarly, the Federal Circuit restated its holding that the therapeutic screening claims were valid. While these holdings were also appealed to the Supreme Court along with the composition of matter claim, the Supreme Court refused to grant *certiorari* on the method patent claim issues.

REQUIREMENTS FOR A PATENT

In order to obtain a patent for a new invention, the invention must be novel, useful, and nonobvious. In addition, the inventor's patent application must describe and enable use of the invention. In fulfilling these requirements, patent examiners typically view genes as nothing more than their chemical structure and use guidelines designed to assist in patenting chemical entities. However, even within that framework, the peculiarities of genes have led to some special rules. This chapter will address each requirement and its application to gene-related patents.

Written Description and Enablement

Section 112 of the Patent Act sets forth the written description and enablement requirements. In terms of the social pact of the patent described above, the written description requirement seeks to assure the public that the inventor had the claimed invention at the time of filing. The enablement requirement ensures that the public will have the information to practice the invention at the end of the patent term. Additionally, taken together, the two requirements serve to limit the reach of a claim into future technologies.[33] While these two requirements work together and the same language often fulfills both requirements, simultaneously describing and enabling, written description and enablement are two distinct requirements for patentability.[34]

Written Description Standard

The purpose of the written description requirement is to "ensure that the scope of the [patent] right . . . does not overreach the scope of the inventor's contribution to the field of the art as described in the patent specification."[35] In order to fulfill that purpose, the courts have developed a standard for written description requiring that a patent application disclosure "reasonably conveys to those skilled in the art that the inventor had possession of the claimed subject matter as of the filing date."[36]

Enablement Standard

Enablement requires that a patent application contain "sufficient information regarding the subject matter of the claims as to enable one skilled in the pertinent art to make and use the claimed invention."[37] Put differently, in order for the invention to be enabled by the disclosure in the patent application, no undue experimentation must be required for one skilled in the art to practice the invention.[38] Enablement does not require the inventor to regurgitate information that is well known in the art.[39] Instead, the skilled practitioner is assumed to have such common knowledge and the inventor must merely teach the practitioner the skills necessary to practice *his* invention.

Patent law recognizes that language is inherently imprecise and that some experimentation may be necessary to completely replicate an invention described in a patent. While some experimentation is allowed, the patent disclosure may not require so much experimentation that the skilled practitioner need reinvent the subject of the patent. To determine how much experimentation is "undue experimentation," the Federal Circuit in *In re Wands* set forth factors to be balanced.[40] Those factors are "(1) the quantity of experimentation necessary; (2) the amount of direction or guidance provided; (3) the presence or absence of working examples; (4) the nature of the invention; (5) the state of the prior art; (6) the relative skill of those in the art; (7) the predictability or unpredictability of the art; and (8) the breadth of the claims."[41]

Written Description and Enablement in Gene-Related Patents

In order for a gene-related invention to be properly described, inventors must not only identify a gene[42] and the protein for which it codes but also describe the function of the protein and the use for the genetic sequence. The Federal Circuit has held that in order to claim a purified gene, the inventors must prove that they understand the chemical structure of the gene being claimed by disclosing the nucleic acid sequence of the gene being claimed.[43] Alternatively, if an inventor chooses to claim a gene in a product-by-process claim, he might "define [the gene] by its method of preparation, its physical or chemical properties, or whatever characteristics sufficiently distinguish it."[44] Generally, however, a gene may not be defined solely by the biological property of encoding for a specific protein[45] or merely by a potential method for isolating it.[46]

Since the chemical structure of the gene, represented by its sequence, is the most concrete way to describe the gene, the USPTO has special rules for the forms of disclosing the nucleic acid sequence of a gene within the patent application.[47] However, an inventor may also describe the gene by placing a specimen, such as a cell line containing the DNA, into a public depository.[48]

Finally, the genetic sequence of a gene from a particular species is not a description sufficient to support a claim to a broader genus. For example, describing the DNA coding for rat insulin would not enable the use of DNA coding for human insulin. Therefore, a specification containing the genetic sequence for rat insulin would be insufficient if the claim were to all mammalian insulin genes.[49]

Enablement of gene-related inventions requires that there be no undue experimentation needed to practice the invention. However, absolute certainty is not required, and the Food and Drug Administration, not the USPTO, is the final arbitrator of effectiveness for human therapeutic use. Therefore, data from in vitro studies approximating in vivo effects are enabling. For example, the in vitro antitumor activity of compounds in a standard murine tumor model was enabling for claims directed at human in vivo effects.[50] Even though more data may be needed, lack of clarity regarding how to use the invention will result in rejection for lack of enablement. For example, a disclosure that does not particularly describe which cyanobacteria can be used as expression vectors, when cyanobacteria vary widely in their abilities, is not enabling.[51] Additionally, overly broad descriptions may defeat enablement as the courts see the inventors as attempting to overreach. Inventors claiming genes with as little as 50 percent homology to a described nucleic sequence[52] or describing a claimed gene as producing a particular protein or a protein that is "biologically functional equivalent thereof"[53] have lost the patent due to lack of enablement.

The Utility Requirement

Three Strands of Utility

The utility requirement for patent procurement arises from several distinct areas. The Constitution itself specifies that patents are to be granted to incentivize progress in the "useful arts."[54] This constitutional directive[55] was deemed so important that a utility requirement was found in the earliest version of the Patent Act in 1790.[56] The utility requirement still exists in today's Patent Act; Section 101 of the Patent Act specifies that the classes of patentable subject matter discussed above must be useful.[57] Additionally, Section 112 of the Patent Act, in discussing the written description and enablement requirements, requires that the *use* of the invention be enabled.[58]

Despite the seemingly straightforward requirement that an invention be useful, courts have struggled with the concept of utility over the history of patent law in the United States. Today the utility requirement encompasses three different types of utility: general (or operational) utility, beneficial (or moral) utility, and specific utility. In spite of the complicated description, all three types of utility are fairly low bars that an invention must hurdle to be entitled to a patent.

Operational utility refers to the requirement that the invention must actually function. Unless the invention falls into the realm of a scientifically dubious field that the patent examiner has reason to believe will not function, the inventor's assertion that his invention has a function generally is not questioned. For example, a machine claimed as a perpetual motion machine would be disallowed without extraordinary proof because such frictionless machines are currently thought to be an impossibility. On the other hand, the same machine claimed to be an amusement or a paperweight would pass the operational threshold because such uses are actual reasonable functions that a machine could have.[59]

Beneficial utility refers to the policy that inventions that have immoral uses should not be patentable. In the past, this doctrine has been used to invalidate patents on inventions such as gambling machines. However, in more recent times, the Federal Circuit has indicated that the morality of such inventions may not be relevant to the invention's patent eligibility.[60] Although U.S. courts have taken a hands-off approach to beneficial utility, some international patent systems still use morality as a benchmark for patentability.

Substantial and specific utility is the most recently developed strand of the utility doctrine. Early on, courts recognized only general and beneficial utility. Judge Joseph Story set the standard by saying that the only utility requirement is that "the invention should not be frivolous or injurious to the well-being, good policy, or sound morals of society."[61] Judge Story reasoned that patents lacking a substantial real-world use would not be commercially successful and would harm no one.[62] Since Judge Story's time, courts began to recognize that granting a patent to an inventor who does not understand the true commercial use for his invention may prevent others from pursuing research on that invention and may further burden an already backed-up patent system.

The Supreme Court embraced this second viewpoint in *Brenner v. Manson* by requiring that inventions must have "specific" and "substantial" uses in order to be patentable.[63] In response, the United States Patent and Trademark Office then issued utility guidance for its patent examiners, requiring that the inventor demonstrate not only credible uses but also specific and substantial uses in order to be granted a patent.[64] Soon thereafter, the Federal Circuit further defined the terms "specific" and "substantial" in *In re Fisher*.[65] Substantial utility requires that the invention disclosure provide some "significant and presently available benefit to the public," and that this benefit must be shown to be available "in its current form, not that it may prove useful at some future date after further research."[66] Specific utility requires disclosure of "a use which is not so vague as to be meaningless," such that the "claimed invention can be used to provide a well-defined and particular benefit to the public."[67]

Utility in Gene-Related Patents

While the bar for the utility requirement is generally low, the utility requirement does play an important role in determining the patentability of many biotechnology inventions, especially gene-related inventions. Mirroring the discussion of general patent laws in the previous section, operational and beneficial utility have little practical impact on the patent eligibility of gene-related inventions. Specific and substantial utility play a far more active role.

Operational utility of gene-related inventions is unlikely to impact patent eligibility. As noted in the previous section, a patent examiner generally accepts an inventor's assertion that the invention works as reported. Moreover, in the biotechnology field, the Food and Drug Administration has the responsibility to determine the efficacy of inventions that concern medical devices or drugs.[68]

Also as noted above, beneficial utility is rarely relied upon to invalidate patents. In the realm of gene-related patents, new technologies such as human and animal cloning, stem cell research, and genetically modified crops have raised ethical issues as to how scientific research should be conducted. However, neither the patent office nor the courts have used beneficial utility to deny patent protection for such technologies. Challenges to the patentability of methods involving embryonic stem cells have failed.[69] Instead, morality-based patent restrictions have been left to congressional discretion, though Congress has been reluctant to impose legal restrictions on technologies still under debate. However, technologies that society at large agrees would be morally reprehensible have been banned. For example, human cloning methods are statutorily exempt from patent protection.[70] The U.S. view of beneficial utility in the biotechnology sector stands in contrast to the views of some international courts. Recently, the European Court of Justice has banned patents protecting stem cell research on ethical grounds.[71]

Specific and substantial utility is likely to be the most important strand of the utility requirement for gene-related inventions. While all gene-related inventions are likely useful for conducting medical research, such utility is not likely to be specific or substantial under patent law.[72] Instead, gene-related patents are likely to require a specific function for the underlying gene product.[73]

In *In re Fisher*, the inventor of several purified nucleic acid sequences, often referred to as expressed sequence tags (ESTs), disclosed seven uses for the ESTs in the patent application—all related to conducting scientific research. Those uses included (1) serving as a molecular marker for mapping the entire maize genome; (2) measuring the level of mRNA in a tissue sample via microarray technology to provide information about gene expressions; (3) providing a source for primers for use in the polymerase chain-reaction process to enable rapid and inexpensive duplication of specific genes; (4) identifying the presence or absence of a polymorphism; (5) isolating promoters via chromosome walking;

(6) controlling protein expression; and (7) locating genetic molecules of other plants and organisms.[74] Applying its definitions of specific and substantial utility, the court concluded that uses for the ESTs disclosed in the patent application were nothing more than research plans for determining the underlying protein, and thus the application disclosed neither specific nor substantial utility. According to the court, no specific utility existed because the disclosed uses were not specific to the claimed ESTs because they could be performed by any EST from any gene.[75] Additionally, the fact that the genes had no known functions in their current form prevented any finding of substantial utility.[76]

The Novelty Requirement

Prior Art Anticipation of an Invention

The novelty requirement exists to ensure that inventions granted patent protection are new to the public, thus increasing the store of human knowledge and encouraging the progress of science and the useful arts. Unlike copyright law, which allows independent creation of a copyrighted work as a defense to infringement, inventions must be absolutely novel in order to be eligible for patent protection. In order to be novel, an invention must not be disclosed by a prior art reference. A prior art reference is a statutorily defined reference that discloses each and every limitation of the claim. The timing of the prior art reference depends on whether a patent application was filed before or after the implementation of the novelty provisions in the America Invents Act (AIA) on March 16, 2013.

The novelty requirement traditionally has been reflected in the U.S. patent system as a rule that, in the case of multiple inventors, the first person to invent receives the patent. However, with the implementation of the novelty provisions in the AIA on March 16, 2013, the United States has moved to a system in which the first inventor to file a patent application with the USPTO will receive the patent, regardless of who first invented. This change to a first-inventor-to-file system reduces the incentive for an inventor to hold onto an invention in secret until finding a commercial demand or until another inventor patents the same invention. It also brings the United States into harmony with most other patent systems worldwide. While the AIA changes the rules for determining which inventor of an invention receives the patent, the basic rules of novelty remain: only one inventor may receive a patent for a given invention, and if the invention was not absolutely novel at the time of the patent applicant's invention, no patent will be allowed.

Specific prior art references for novelty and statutory bars are defined by Section 102 of the Patent Act. For example, for patent applications filed before March 16, 2013, novelty requires that, *before the date of invention*, the invention must not be known or used by people in the United States other than the inventor and must not be patented or described in a printed publication anywhere in the world.[77] Any evidence of knowledge, use, patenting, or a printed publication of the invention would erase its novelty, making it ineligible for patent protection. For patent applications filed after the implementation of the novelty provisions of the AIA, the analogous prior art definitions state that the invention must not be patented or described in a printed publication or otherwise available to the public *before the date of filing* the patent application of the invention.

In order for a prior art reference to anticipate the invention and render it non-novel, the prior art reference must disclose every element of the claimed invention, either explicitly or inherently. All elements of the invention must be in one single prior art reference. Explicit anticipation is fairly easy to understand—all the elements of the invention must be explicitly disclosed in the prior art reference. Inherent anticipation occurs when the prior art reference does not explicitly disclose every element but the missing elements would necessarily be present, even if not expressly stated.[78] That inherent anticipation is allowed means that merely discovering a new property of a material does not render a material, formerly in the prior art, suddenly novel and patentable.[79]

Additionally, the reference must be enabling. This enablement standard for prior art references is slightly different from the enablement requirement previously discussed for patents. Here, the prior art reference must enable only making the invention while enablement for patentability must enable both making and using the invention.[80]

Anticipation of Gene-Related Inventions

The pace of advancement in genetic research makes inherent anticipation very important to inventors of gene-related inventions. For example, a prior art plasmid that had not been previously sequenced inherently anticipated a promoter sequence from that plasmid because the plasmid necessarily contained the same DNA sequence as the promoter.[81] It did not matter that no one had previously recognized or uncovered the promoter sequence. Prior to sequencing, that promoter sequence had existed in the plasmid and thus kept the promoter from being novel. Similarly, a prior art reference disclosing a genetic mutation and a method of testing for the mutation, absent any correlation with disease, anticipated the later invention of a method for testing for a disease by using the disclosed method to screen for the genetic mutation.[82]

The high skill level of bioscience researchers also plays a role in inherent anticipation of gene products. For example, a reference that lists every 15-base *sense* oligodeoxynucleotide in a known nucleic acid sequence may anticipate specific *antisense* sequences based on the disclosed sense sequence. Remembering that the prior reference must be enabling, the sense sequences anticipate the antisense sequences because a typical scientist in the field could make the antisense sequences from the referenced sense sequences without any more information.[83] Therefore, the sense sequences themselves provide all the information necessary to make the antisense sequences, and thus the antisense sequences are not novel.

Many stakeholders in the biosciences do not support the patenting of genes in the fear that too many patents on such basic concepts will block research and drug development. These stakeholders include both university scientists and large pharmaceutical companies. The ability for published sequence lists to anticipate gene patents has led many to mount efforts to put as much information into the public domain as possible, so that the information can act to anticipate and block gene patents.

As a result of these disclosures, databases such as those mentioned in Chapter 12 are often rich sources of prior art references. For example, the Human Genome Project's GenBank[84] is an annotated genetic sequence database containing all publicly available DNA sequences. The National Institutes of Health requests that scientists disclose information as it becomes available, but it has policies in place to respect publication and property rights of the researchers. Single nucleotide polymorphisms (SNPs), thought to be one of the next big research tools in identifying the genetic bases of many diseases and developing treatments, have been mapped and are publicly available.[85] The mapping was completed using both federal funding as well as through the work of the SNP Consortium, a consortium of the U.K. Wellcome Trust and ten pharmaceutical companies. This publicly available SNP information is now the basis of several bioinformatics projects looking at population level variation.[86]

The Nonobviousness Requirement

Obviousness Generally

While novelty is concerned with whether a particular invention has been previously invented, a nonobviousness requirement is directed at minor improvements over the prior art; both of these requirements must be met in order for an invention to be eligible for a patent.[87] This nonobviousness requirement is a recognition that a novel invention may be such a small improvement over prior inventions that it may not advance the field enough to warrant the grant of a patent. There is no societal advantage to trading patent rights for an invention so trivial in advancement that it would have been invented numerous times anyway because it was obvious to those skilled in the field.

Anticipation can occur only when one prior art reference fully discloses the claimed invention, but an invention may be deemed to be obvious based on the use of multiple prior art references that can be read together to encompass all of the claimed elements. Also, unlike anticipation, a nonobviousness analysis has a perspective—the obviousness is through the eyes of the average person of skill in the field of the invention. This hypothetical person is often called the PHOSITA, an acronym for "a person having ordinary skill in the art." Due to the ability to combine prior art references and the use of the PHOSITA as a reference frame, nonobviousness is often a tougher hurdle than is novelty.

The fundamental question to a nonobviousness analysis for an invention is whether a PHOSITA, armed with the knowledge available at the time the invention was made, would have used that knowledge to come to the same conclusion. Think back to the pencil patent example. It may well be that combining the eraser and the pencil into one unit would have been obvious to a PHOSITA of the time, and thus that rather trivial innovation would not have qualified for a patent. But how do we tell whether something like our pencil/eraser combination was an obvious innovation? The basic test for nonobviousness was defined by the Supreme Court decades ago in *Graham v. John Deere Co.*[88]

The analytical framework set forth by the Supreme Court involves four steps. First, the court must determine the scope of content of the prior art. The categories of prior art are basically the same as the categories of prior art for anticipation. The second step requires the court to determine the differences between the prior art and the claimed invention at issue. Third, the level of ordinary skill in the field of the invention must be determined. The fourth step is to use the determinations of the first three steps and establish whether the claimed invention would have been obvious to a PHOSITA at the time the invention was made.

In making this final determination of nonobviousness, courts can consider "secondary considerations," such as whether the prior art teaches away—i.e., whether the prior art suggests that the claimed invention is not a potential solution.[89] Other secondary considerations include the commercial success of the invention and the failure of others in trying to solve the same problem addressed by the invention.[90]

A critical determination in this nonobviousness analysis is the identity and characteristics of the PHOSITA. For instance, a court may determine that the level of ordinary skill for a biotechnology invention may be that of a Ph.D.-level life sciences researcher, but for an invention to impede drafts from under doors it may be that of a high-school graduate. The Supreme Court has recently held that the person of average skill in the art also has average background knowledge and average creativity.[91] This essentially means that routine advances over the prior art are unlikely to be considered nonobvious by a court. As stated by the Supreme Court, "the results of ordinary innovation are not the subject of exclusive rights under the patent laws. Were it otherwise patents might stifle, rather than promote, the progress of useful arts."[92]

Obviousness in Gene-Related Patents

Perhaps one of the tests most applicable for use with gene and gene-related patents is the "obvious-to-try" test for nonobviousness. In order for the invention to be obvious, the development of the invention must have included a finite number of identified, predictable solutions that resulted from the predictable use of prior art elements according to their established functions.[93] A finding of obviousness under the obvious-to-try standard does not require absolute certainty of success, but merely a reasonable expectation of success.[94]

As advancements in technology make routine work out of once-extraordinary methods and discoveries, the potential to be obvious rises for inventions developed using those now-standard techniques. Today, standard molecular techniques to develop a DNA sequence from a known protein sequence are highly likely to make any invention of the resultant DNA obvious.[95] Once the protein sequence is known, it is relatively straightforward to identify the corresponding DNA sequence from the possible sequences. As long as our PHOSITA could pursue a definite set of potential solutions with a reasonable expectation of success, the resulting DNA sequence would be obvious in light of the protein. Such a result would not have been possible before advances in technology and prior cases held that the gene sequence would be nonobvious.

Inventions in which the obvious-to-try test is not applicable due to unpredictable results or large numbers of potential solutions are more likely to be nonobvious. Another way to circumvent an obviousness determination is to patent a method of using the gene rather than or in addition to patenting the gene itself. For instance, patenting a method of using the gene to conduct research on a particular therapy or patenting a method of using the gene as a target for a gene therapy would protect the use of the gene that the inventor plans to monetize. Since all genes must have a utility in order to qualify for patent protection, framing the patent in terms of use is relatively easy and gives the inventor some insurance against the patent being denied or invalidated due to obviousness.

CONCLUSION

Gene-related patents, whether directed at genetic sequences or methods of use, are an important consideration in planning any bioinformatics research project. The state of the law on gene patenting is not always clear, and will evolve as our understanding of and skill in the genetic arts progresses. The patent debate is complicated by the intimate connection that many feel toward the subject matter in question. In the meantime, the courts attempt to use the line between incentives for inventors and public access to material vital to industry development as guidance in determining when to grant patent protection.

NOTES

1. U.S. Const. art I, § 8, cl. 8.
2. The other types of patents are plant patents and design patents.
3. 35 U.S.C. §§ 101 *et seq*.
4. 35 U.S.C. § 271(a).
5. In 2011, the average pendency of a utility patent application that matured in to a patent was about 34 months. U.S. Patent Statistics Chart, http://www.uspto.gov/web/offices/ac/ido/oeip/taf/us_stat.htm.
6. The USPTO routinely publishes statistics on the number of patent applications received and the number of grants of patents made. These numbers do not expressly coordinate, because applications are rarely granted in the same year they are received. *Id.*
7. 35 U.S.C. § 154(a)(2).

8. The real patent for a pencil/eraser combination was issued to Hymen Lipman in 1858. U.S. Patent No. 19,783 (issued Mar. 30, 1858).

9. ORGANIZATION FOR ECON. CO-OPERATION & DEV., GENETIC INVENTIONS, INTELLECTUAL PROPERTY RIGHTS AND LICENSING PRACTICES: EVIDENCE AND POLICES 15 (2002).

10. Richard A. Epstien & Bruce N. Juhlik, *Is There a Biomedical Anticommons?*, 27 REG. 54 (2004).

11. 35 U.S.C. § 101.

12. *See* Bilski v. Kappos, 130 S. Ct. 3218, 3225 (2010) ("[i]n choosing such expansive terms . . . modified by the comprehensive 'any,' Congress plainly contemplated that the patent laws would be given wide scope.").

13. 447 U.S. 303 (1980).

14. *Id.* at 309; *see also* Gottschalk v. Benson, 409 U.S. 63, 67 (1972) ("Phenomena of nature, though just discovered, mental processes, and abstract intellectual concepts are not patentable, as they are the basic tools of scientific and technological work.").

15. *Bilski*, 130 S. Ct. at 3225 (quoting Funk Bros. Seed Co. v. Kalo Inoculant Co., 333 U.S. 127, 130 (1948)).

16. *Id.* at 3230 (quoting Diamond v. Diehr, 450 U.S. 175 (1980)).

17. 132 S. Ct. 1289 (2012).

18. U.S. Patent No. 6,355,623 (issued Mar. 12, 2002); U.S. Patent No. 6,680,302 (issued Jan. 20, 2004).

19. *Prometheus*, 132 S. Ct. at 1305 ("[W]e conclude that the patent claims at issue here effectively claim the underlying laws of nature themselves. The claims are consequently invalid.").

20. *Id.* at 1291.

21. *Id.* at 1298.

22. 66 C.F.R. § 1093.

23. Among these patents were the patents involved in the case, including U.S. Patent No. 5,747,282 (issued May 5, 1998), U.S. Patent No. 5,837,492 (issued Nov. 17, 1998), U.S. Patent No. 5,693,473 (issued Dec. 2, 1997), U.S. Patent No. 5,709,999 (issued Jan. 20, 1998), U.S. Patent No. 5,710,001 (issued Jan. 20, 1998), U.S. Patent No. 5,753,441 (issued May 19, 1998), and U.S. Patent No. 6,033,857 (issued Mar. 7, 2000).

24. U.S. Patent No. 5,747,282 (issued May 5, 1998).

25. Representative method claims include claim 1 of the '999 and '001 patents:

> A method for detecting a germline alteration in a BRCA1 gene, said alteration selected from the group consisting of the alterations set forth in Tables 12A, 14, 18 or 19 in a human which comprises *analyzing* a sequence of a BRCA1 gene or BRCA1 RNA from a human sample or *analyzing* a sequence of BRCA1 cDNA made from mRNA from said human sample with the proviso that said germline alteration is not a deletion of 4 nucleotides corresponding to base numbers 4184–4187 of SEQ ID NO: 1. '999 patent claim 1 (emphases added).
>
> A method for screening a tumor sample from a human subject for a somatic alteration in a BRCA1 gene in said tumor which comprises [ ] *comparing* a first sequence selected from the group consisting of a BRCA1 gene from said tumor sample, BRCA1 RNA from said tumor sample and BRCA1 cDNA made from mRNA from said tumor sample with a second sequence selected from the group consisting of BRCA1 gene from a nontumor sample of said subject, BRCA1 RNA from said nontumor sample and BRCA1 cDNA made from mRNA from said nontumor sample, wherein a difference in the sequence of the BRCA1 gene, BRCA1 RNA or BRCA1 cDNA from said tumor sample from the sequence of the BRCA1 gene, BRCA1 RNA or BRCA1 cDNA from said nontumor sample indicates a somatic alteration in the BRCA1 gene in said tumor sample. '001 patent claim 1 (emphasis added).

Ass'n for Molecular Pathology v. USPTO, 653 F.3d 1329, 1334–35 (Fed. Cir. 2011).

26. Ass'n for Molecular Pathology v. USPTO, 702 F. Supp. 2d 181 (S.D.N.Y. 2010).

27. Ass'n for Molecular Pathology v. USPTO, 653 F.3d 1329 (Fed. Cir. 2011).

28. "Applying this test to the isolated DNAs in this case, we conclude that the challenged claims are drawn to patentable subject matter because the claims cover molecules that are markedly different—have a distinctive chemical identity and nature—from molecules that exist in nature." Ass'n for Molecular Pathology v. USPTO, 689 F.3d 1303, 1328 (Fed. Cir. 2012), *cert. granted in part*, 133 S. Ct. 694 (2012).

29. Mayo Collaborative Servs. v. Prometheus Labs., Inc., 132 S. Ct. 1289, 1291 (2012).

30. *Supra* note 25.

31. "We conclude that Myriad's claims to 'comparing' or 'analyzing' two gene sequences fall outside the scope of § 101 because they claim only abstract mental processes." *Ass'n for Molecular Pathology*, 653 F.3d at 1355.

32. *Id.* at 1303.

33. Kevin Collins, *An Initial Comment on Ariad: Written Description and the Baseline of Patent Protection for After-Arising Technology*, 24 PATENTLY-O PATENT L.J. (2010).

34. Ariad Pharma. v. Eli Lilly & Co., 598 F.3d 1336 (Fed. Cir. 2010).

35. Univ. of Rochester v. G.D. Searle & Co., 358 F.3d 916, 920 (Fed. Cir. 2004) (quoting Reiffen v. Microsoft Corp., 214 F.3d 1342, 1345 (Fed. Cir. 2000)).

36. *Ariad*, 598 F.3d at 1351 (internal citations omitted).

37. *Manual of Patent Examiner's Procedures: 2164.01 Test of Enablement, available at* http://www.uspto.gov/web/offices/pac/mpep/s2164.html#d0e215268 (last visited Nov. 5, 2012).

38. Mineral Separation v. Hyde, 242 U.S. 261, 270 (1916); *In re* Wands, 858 F.2d 731, 737 (Fed. Cir. 1988).

39. *Manual of Patent Examiner's Procedures: 2164.01 Test of Enablement, available at* http://www.uspto.gov/web/offices/pac/mpep/s2164.html#d0e215268 (last visited Nov. 5, 2012).

40. *Wands*, 858 F.2d at 731.

41. *Id.*

42. Fiers v. Sugano, 25 U.S.P.Q. 2d (BNA) 1601 (Fed. Cir. 1993).

43. Univ. of Cal. v. Eli Lilly & Co., 119 F.3d 1559 (Fed. Cir. 1997); Amgen Inc. v. Chugai Pharma. Cal. Co., 927 F.2d 1200 (Fed. Cir. 1991).

44. Amgen, 927 F.2d at 1200.

45. *Id.* at 1206.

46. Fiers v. Sugano, U.S.P.Q. 2d (BNA) 1601 (Fed. Cir. 1993).

47. 37 C.F.R. § 1.821.

48. Enzo Biochem, Inc. v. Gen-Probe, Inc., 323 F.3d 956 (Fed. Cir. 2002).

49. Univ. of Cal. v. Eli Lilly & Co., 43 U.S.P.Q. 2d (BNA) 1398 (Fed. Cir. 1997).

50. *In re* Brana, 51 F.3d 1560 (Fed. Cir. 1995).

51. *In re* Vaeck, 947 F.2d 488 (Fed. Cir. 1991).

52. *Ex parte* Ishizaka, No. 91-2539, 1992 WL 336794 (B.P.A.I. July 14, 1992).

53. *Ex parte* Maizel, No. 91-2301, 1992 WL 519152 (B.P.A.I. Apr. 23, 1993).

54. Congress has the authority "[t]o promote the Progress of Science and useful Arts, by securing for limited Times to . . . Inventors the exclusive Right to their . . . Discoveries." U.S. Const. art I, § 8, cl. 8.

55. Section "101 serves to ensure that patents are granted on only those inventions that are 'useful.' This second purpose has a Constitutional footing—Article I, Section 8." *Manual of Patent Examiner's Procedures: 2107.01 General Principles Governing Utility Rejections, available at* http://www.uspto.gov/web/offices/pac/mpep/s2107.html (last visited Nov. 5, 2012).

56. "[H]ath or have invented or discovered any *useful* art, manufacture, engine, machine, or device, or any improvement therein not before known or used, and . . . the Secretary of State, the Secretary for the department of war, and the Attorney General, or any two of them, if they shall deem the invention or discovery *sufficiently useful* and important." Patent Act of 1790, ch. 7 § 1, 1 Stat. 109-112 (1790) (current version at Title 35, United States Code).

57. "Whoever invents or discovers any new and *useful* process, machine, manufacture, or composition of matter, or any new and *useful* improvement thereof, may obtain a patent therefor, subject to the conditions and requirements of this title." 35 U.S.C. § 101.

58. *Id.* § 112.

59. But the invention might still fail the substantial utility requirement.

60. Juicy Whip, Inc. v. Orange Bang, Inc., 51 U.S.P.Q. 2d (BNA) 1700 (Fed. Cir. Aug. 6, 1999).

61. Lowell v. Lewis, 15 F. Cas. 1018, 1019 (C.C.D. Mass. 1817).

62. "[W]hether it be more or less useful is a circumstance very material to the interests of the patentee, but of no importance to the public. If it be not extensively useful, it will silently sink into contempt and disregard." *Id.*

63. Brenner v. Manson, 383 U.S. 519 (1966).

64. http://www.gpo.gov/fdsys/pkg/FR-2001-01-05/html/01-322.htm.

65. *In re* Fisher, 421 F.3d 1371 (Fed. Cir. 2005).

66. *Id.*

67. *Id.*

68. *In re* Brana, 51 F.3d 1560 (Fed. Cir. 1995).

69. *U.S. office upholds embryonic stem cell patents*, JSOnline (June 27, 2008), http://www.jsonline.com/business/29551579.html.

70. Section 33(a) of the Leahy-Smith America Invents Act provides: "Notwithstanding any other provision of law, no patent may issue on a claim directed to or encompassing a human organism."

71. Ewen Callaway, *European Court Bans Patents Based on Embryonic Stem Cells*, Nature, www.nature.com/news/2011/111018/full/news.2011.597.html (last visited Nov. 18, 2012).

72. *In re* Fisher, 421 F.3d 1374 (Fed. Cir. 2005).

73. *Id.*

74. *Id.*

75. *Id.*

76. *Id.*

77. 35 U.S.C. § 102(a).

78. Schering Corp. v. Geneva Pharma., 339 F.3d 1377 (Fed. Cir. 2003).

79. Atlas Powder Co. v. Ireco Inc., 190 F.3d 1342, 1347 (Fed. Cir. 1999).

80. Rasmusson v. SmithKline Beecham Corp., 413 F.3d 1318, 1325–26 (Fed. Cir. 2005) (citing *In re* Schoenwald, 964 F.2d 1122, 1124 (Fed. Cir. 1992); *In re* Donohue, 632 F.2d 123, 126 n.6 (C.C.P.A. 1980)) ("[A] prior art reference need not demonstrate utility in order to serve as an anticipating reference under section 102.").

81. *In re* Crish, 393 F.3d 1253, 1258 (Fed. Cir. 2004).

82. Billups-Rothenberg, Inc. v. ARUP, 642 F.3d 1031 (Fed. Cir. 2011).

83. *In re* Gleave, 560 F.3d 1331 (Fed. Cir. 2009).

84. GenBank, http://www.ncbi.nlm.nih.gov/genbank/ (last visited Nov. 5, 2012).

85. dbSNP Short Genetic Variations, http://www.ncbi.nlm.nih.gov/SNP/ (last visited Nov. 5, 2012).

86. Human Genome Project Information, http://www.ornl.gov/sci/techresources/Human_Genome/faq/snps.shtml (last visited Nov. 5, 2012).

87. 35 U.S.C. § 103(a) ("A patent may not be obtained . . . if the differences between the subject matter sought to be patented and the prior art are such that the subject matter as a whole would have been obvious at the time the invention was made to a person having ordinary skill in the art to which said subject matter pertains.").

88. 383 U.S. 1 (1966).

89. *In re* Gurley, 27 F.3d 551, 553 (Fed. Cir. 1994) ("[A] person of ordinary skill, upon examining the reference would be discouraged from following the path set out in the reference, or would be led in a direction different from the path that was taken by the applicant.").

90. Graham v. John Deere Co., 383 U.S. 1 (1966).

91. "A person of ordinary skill is also a person of ordinary creativity, not an automaton." KSR Int'l Co. v. Teleflex Inc., 550 U.S. 398, 421 (2007).

92. *Id.* at 427.

93. *In re* Kubin, 561 F.3d 1351, 1360 (Fed. Cir. 2009) (citing KSR Int'l Co. v. Teleflex Inc., 550 U.S. 398, 417, 421 (2007)).

94. *Id.* at 1360 (quoting *In re* O'Farrell, 853 F.2d 894, 903–04 (Fed. Cir. 1988)).

95. *Id.*

CHAPTER 7

Privacy, Property Rights, and Gene Ownership

Richard A. Spinello

INTRODUCTION

In his remarkable book about the history of information, James Gleick explains that just as we now consider the universe as a cosmic computer, the body too must be regarded as an information processor. It is no surprise, therefore, that the field of genetics has flourished in ever-closer parallel with the elaboration of information theory. When DNA was discovered, we realized that the encoded information it contains represents its proper function: to direct the development and growth of the human organism. Every living being is determined by this code, a long number written in a four-digital system. Thus, the field of genetics is "the language of cells."[1] In order to properly understand this language and the precise role of DNA, we must begin by talking about those cells.

Cells are the basic units of all living organisms. Within cells are the nuclei or life force of the cells. DNA and RNA are the nucleic acids found in an organism's cells, and DNA is the molecule that stores genetic material. Within the nucleus of every cell is about six feet of DNA. DNA is composed of genes, which are really "stretches" or strains of that DNA. Genes are organized into chromosomes and it is through chromosomes that genetic information is transmitted. A human's chromosomes contain approximately 30,000 genes, and this complete set of genes is known as the human genome.[2] These genes contain the biological information necessary for making certain proteins. In effect, each gene is analogous to a sentence with a four-letter alphabet, A, T, C, and G (representing the nucleotide bases that form the genes: adenine, thymine, cytosine, guanine), which combine in pairs to communicate with the cell and instruct its development in certain ways. It is this genetic information, for example, that instructs cells to make black hair instead of brown hair.[3]

The Human Genome Project mapped and sequenced these genes. This effort has enabled genetic testing and also opened the door for gene therapy. Through a blood test or tissue sample, it is possible to determine aspects of an individual's genetic status, which, in conjunction with the human genome map, will allow doctors to determine whether an individual has defective genes that may predispose that individual to a certain illness or other debility. For example, researchers have isolated two genes, BRCA1 and BRCA2, which function to suppress breast tumors. When a problem or mutation occurs with either of these genes, breast cancer can be the result. Once genetic diseases have been diagnosed, the goal is to develop therapies that correct the mutation. Consider the disease known as phenylketonuria, which is triggered by a mutation to a gene that breaks down the molecule called phenylalanine; phenylalanine can cause brain damage if it builds up in the bloodstream. The optimal cure is to repair this defective gene so that the person's metabolism is restored to a normal state.[4]

Each person's genetic information is extremely valuable, since genetic tests can be useful predictors of one's future health prospects. But this information can also be detrimental if it is misappropriated or misused. It can be manipulated by businesses, governments, or even scientific researchers in ways that could potentially compromise a subject's well-being. Also, this information can be the basis for developing cures or diagnostic testing methods that can yield high returns for a biotech company or other enterprises. Henrietta Lacks was a poor black woman who died of cervical cancer in the 1950s, eight months after her diagnosis. With the help of this woman's cancer cells, researchers developed a method for reproducing human cells in the lab. Mrs. Lacks's cells became the first human cells to reproduce prolifically in a laboratory setting, and scientists have reproduced some 50 million metric tons of her cells over time. The so-called HeLa cell line has contributed to major developments in the field of life sciences, including the polio vaccine, cancer treatments, and in vitro fertilization. Unfortunately, the woman's family received no compensation from those who profited from this stream of research on her cells.

Breakthroughs in genetics have also given rise to the expanding field of population genomics research. In one well-known population project, deCODE Genetics was given exclusive research access to the genetic data of Iceland's citizens. Iceland's Healthcare Database links together separate databases of medical records, genealogical records, and genetic information based on DNA samples received voluntarily from over 70,000 citizens. deCODE's choice of Iceland was based on the genetic homogeneity and small size of its population. The goal was to isolate the genetic mutations between the DNA of those carrying certain diseases and the DNA of the general population without the disease. deCODE has already identified genes associated with maladies such as schizophrenia, Parkinson's disease, and type 2 diabetes.

Despite its promising results, however, this project has been sharply criticized because of inadequate measures to fully protect privacy rights. Although donors of genetic information gave their "informed consent" prior to the collection of their DNA, they were not informed that this genetic data would be cross-referenced with other medical data in the Iceland study. In addition, deCODE has property rights in the proteins or DNA segments isolated from those blood and tissue samples. Hence any financial value from this project accrues to deCODE and to the Iceland government, but not to those who contributed their tissue samples that became the basis of this research.

Thus, there are two powerful ethical issues that revolve around advances in genetics as made evident in the Lacks and deCODE cases: how is privacy about one's genetic information to be amply protected, and who owns tissues extracted from the body along with DNA segments isolated from those samples? These two issues are intimately interconnected. Privacy requires control, and proprietary rights may be the optimal means of providing such control. However, as we will demonstrate, the assignment of proprietary rights is not without serious disadvantages, and hence the benefits do not appear to outweigh the costs. Those same social costs arise when companies are granted gene patents for their discoveries. Exclusive ownership rights thwart research efforts and are inconsistent with the normative rationale for justifying property rights. We begin with a deeper analysis of the privacy issue as a springboard for the main discussion on the theme of intellectual property rights and gene patents.

PRIVACY RIGHTS AND GENETIC INFORMATION

The confidentiality of genetic or genomic information is fundamental for the protection of a person's overall medical privacy. While standard measures to protect privacy such as secure databases and informed consent seem adequate enough, certain problem areas demand attention. For example, people might voluntarily submit to researchers or private companies blood or tissue samples from which their DNA can be extracted. Often the purpose is for the sake of genetic testing that probes for certain mutations. With the donor's permission, this identifiable data may be archived and then made anonymous so it can be used for research purposes. Research protocols typically call for a correlation with other data sets. Disclosure risk may seem low since the data is tightly secured and made anonymous. Nonetheless, evidence shows that the probability of disclosure is fairly high because the correlation and cross-referencing with other databases can often lead to the revelation of a person's identity. The full data environment for a genetic or bio-database typically includes census data or even lifestyle databases. The correlation of these data sets may indirectly reveal one's identity. This is especially easy in the case of "special uniques," individuals in a given subset of the population with unique traits. Thus, a 30-year-old, recently widowed woman with five children living in a small city may have her genetic disorder revealed when her anonymous genetic and personal data (given willingly to a medical researcher) is linked to census data for this area.[5]

Unfortunately, there are plenty of "data intruders," third parties who might seek out this type of information, which can be used for research, forensic, or marketing purposes. When identifiable genetic information becomes too widely available or falls into the wrong hands, this infringement of privacy rights could have injurious effects, including discrimination and stigmatization. Insurers, for example, could use this information to deny specific coverage or increase rates. Employers might rely on the information to discriminate in their hiring and promotion practices. Discrimination can assume many other guises. It has been suggested that women who have the BRCA gene mutation should reveal it to their prospective husbands so they can find a mate with "better genes," if they are so inclined.[6] It is certainly not inconceivable that the disclosure of genetic information may factor into major lifestyle decisions in the not-too-distant future. Some countries have legislation that protects against discrimination, but those protections are limited and difficult to enforce.

The other problem is that data mining and computations assign these nonidentifiable subjects to different profile groups, usually without their permission or knowledge. Since the data has been made anonymous, the data subject loses all control even under strict privacy laws such as the European Union directive. But group profiles often become the basis for policy decisions, and individuals could be judged and injured in this process. Perhaps a group profile consists of women with a fairly common but serious genetic mutation who have given birth to five or more children. If interested third parties discern that there is a greater incidence of this genetic illness among women belonging to this group, women with five or more children may find themselves as the objects of subtle forms of discrimination because it is assumed that they have a higher risk of contracting this debilitating illness. With this in mind, Vedder calls for "categorial privacy" as well as individual privacy that will protect people both as individuals and as members of groups even when their identity has been initially stripped away. Obviously, infringements of categorial privacy based on stereotyping can yield the same negative consequences as infringements of individual informational privacy. The only remedy is a more "careful assessment" of the purposes for which group profiles are used.[7]

Some privacy advocates are convinced that the unwarranted release of genetic information is so harmful that they support "genetic exceptionalism." According to this principle, genetic information is qualitatively different from other forms of health-related information, and therefore it warrants special protection. The exceptionalists argue that given its highly sensitive nature, genetic information must be regulated more stringently than other forms of personal data.[8]

The principle of genetic exceptionalism seems plausible enough. Genetic information is unique in that it reveals not only one's present medical condition but also predispositions to certain ailments such as diabetes or cancer. One's DNA is analogous to a "diary" of a person's future well-being. More importantly, genetic information provides data not just about a single individual from whom a DNA sample was taken but about his or her family members. As O'Neill points out, this data is "intrinsically familial rather than individual."[9] What is revealed about a person's own genetic makeup reveals something about the genetic makeup of that person's family members as well, especially parents, siblings, and children. The deliberate or inadvertent divulgence of this data could adversely affect family dynamics.

Anti-exceptionalists argue, however, that there is nothing intrinsically different about this data that warrants special normative or legal consideration. All medical data is highly sensitive and deserves an elevated level of protection. We cannot resolve the debate over exceptionalism in this chapter but we can at least concur with O'Neill's thesis that the use of personal genetic data is distinctive because third parties with no personal interest in this data are keen to gain access to it.[10] The creation of genetic profiles and the interpretation of genetic data as a predictor of one's health, along with the scope and intensity of data mining of this information, can create what Tavani calls "unanticipated outcomes."[11] In addition, given that this data reveals information not just about the donor but about his or her family, it is difficult to construct a valid informed consent structure that is both comprehensive and transparent. Hence there is need to somehow ensure that access to an individual's genetic data will remain restricted.

In recognition of the dangers of confidentiality breaches, some countries, including the United States, have enacted federal laws to protect genetic privacy. In 2008 the U.S. Congress enacted into law the Genetic Information Nondiscrimination Act (GINA). This law prohibits the use of genetic information to deny someone health insurance, and it penalizes employers who use this information in making decisions about hiring and firing employees. While this is a good first step, it is questionable whether this legislation goes far enough. Many other forms of potential discrimination based on genetic data are not covered by the law. The government itself could use genetic information to profile people who might be future criminal offenders based on their DNA or, where there is socialized medicine, as a basis to deny expensive treatment. For this reason, critics of GINA call for more extensive or "thicker" privacy safeguards.

But how could such safeguards be implemented? The issue is complicated because of the ambiguity regarding the ownership status of genetic information. Who owns the human tissues extracted from the body and the DNA information they contain? Some of those who support the principle of genetic exceptionalism contend that a proprietary right to one's genetic information would be the most viable means of securing the confidentiality of that information.[12] Ownership was the central issue in *Moore v. Regents of the University of California*. John Moore filed suit against researchers at the university who patented a cell line from the tissues derived from his diseased spleen once it was surgically removed from his body. The California court ruled against Moore's claim of any proprietary right over this genetic material, but in a dissenting opinion Judge Stanley Mosk declared that "the patient has a legally protected property interest in his own body and its products."[13]

This notion has even surfaced in one significant piece of state legislation. Oregon's Genetic Privacy Act included such a "property clause": "an individual's genetic information is the property of the individual." Oregon believed that such a statute would provide unambiguous guidance to courts grappling with these questions. The property clause was ultimately removed from the bill, much to the dismay of some privacy advocates, who were quick to find fault with Oregon's reversal: "[B]y removing the property interest that had been given to individuals with regard to genetic information, Oregon has decreased the strength of privacy associated with their genes."[14] McLochlin suggests that Oregon was on the right track and capitulated too easily to the demands of the scientific community and biotech companies. The assumption of McLochlin and those who crafted the law was that a proprietary right would be a highly efficacious means of preserving one's privacy. As Miller has indicated, a "property right vested in the subject" is the surest way to give someone autonomous control over that information, whether it be financial data or one's genetic code.[15]

Similarly, Koepsell has argued for the need to have greater personal control over our tissue and blood samples and over our "individual genomes"—that is, the portion of our genome unique to us, including those genetic mutations that can trigger various disorders. He envisions this control taking the form of "positive laws that *create new rights* over our individual genes" such that the individual becomes a proprietor over his individual genetic identity.[16] In this way our privacy will be more securely protected, and the possessor of this genetic information could strike a bargain with the scientific community for the use of his or her genes. This approach would "create a marketplace with more equality for the bargaining parties."[17]

Whether a property right in this context would provide the security envisioned by the Oregon state legislature is somewhat dubious, given the asymmetries between the information donors and the researchers who collect this information. But this debate raises a larger question about the normative justification for a person's proprietary interest in his or her extracorporeal bodily material, along with specific genetic information derived from that material. Researchers have fiduciary duties to safeguard this information and to seek consent for its use, but property rights would undoubtedly provide more potent legal remedies and sanctions when cells or genetic data are misused. Moreover, applying property rights to this data will make it easier for individual "owners" to share in the rewards when their genetic material is later commercialized. But what are the limitations or harmful side effects of resolving privacy rights and ensuring distributive justice by means of a new set of property rights? While the property paradigm has a certain appeal, we contend that it is morally inexpedient because these property rights will yield a negative social impact: they will further encourage the body's commodification and obstruct research essential for the common good.

COMMODIFYING THE HUMAN BODY

It would seem logical that the law should recognize a person's ownership of his cells along with the DNA sequences encoded in those cells (even in their state outside the body). Ownership would give an individual more control over this sensitive information and presumably prevent the apparent injustice that befell Moore and the Lacks family, since it would enhance the chances of remuneration for the diagnostic tests or therapies developed from these cells or DNA.

But is it prudent to regard the body and its constituent parts as a form of legal property? What would be the unintended consequences of such an ethical and legal strategy? To begin with, endowing bodily material with this sort of property right transforms it into a commercial product and thereby thrusts it into the sphere of market competition. As a result, the person is no longer someone who transcends marketplace transactions. Instead, market culture encroaches on the sacred domain of the person. Property rights in extracorporeal body parts and DNA will contribute to a commodification trend that is already well underway within the scientific milieu. The general tendency to commodify the body and treat it as a product has already manifested itself in research on embryos and the destruction of embryos to harvest pluripotent stem cells. That research is supported by the sale of human eggs, which are needed for the artificial creation of those research embryos. The creation of certain forms of cyborgs along with innovative biotechnologies such as cloning and genetic engineering that manipulate human material also reflect this trajectory toward broader commodification. The question of which sorts of manipulation are ever morally permissible is beyond the scope of this book. But the cumulative effect of these technologies is to encourage the body's objectification, that is, its treatment as the discrete "object" of medical research or scientific exploitation. The body is polarized from the personal subject such that its use as a scientific experiment is not regarded as transgressive of personhood but merely as another form of conquest. We come to see ourselves as both "subject and object . . . literally creatable through genetic engineering," but in so doing we lose a proper appreciation of our dignity.[18]

This tendency toward objectification has its roots in the same philosophical soil that has nurtured the Cartesian tradition of subjectivism, such that objects become conceivable only if they are controlled by the subject. According to Heidegger, subjectivism comes to manifest itself in the prevailing attitude of technicity (*Technik*). Heidegger's term expresses how all beings, even the person and his or her body, are raw material for our use and disposal. Beings are submitted to man's tight control, objectified by his calculations (*Berechnung*), planning (*Planung*), and cultivation (*Züchtung*). This attitude underlies the relentless drive of technology that results in its unequivocal domination in modern society where everything is reduced to a pawn that must submit to the subject's designs. Within this worldview it is not surprising that the value of the objectified body is linked to its economic usefulness for the purpose of scientific progress.[19]

In legal circles the famous 1980 *Diamond v. Chakrabarty* case seems to have set a broad legal precedent that provides some impetus for this commodification trend by allowing the patenting of living organisms. In that case, the judges opined that the patent statute should cover "anything under the sun that is made by man." The Court ruled that genetically altered life forms such as plants and animals could be patented. This ruling in conjunction with the *Moore* case opened the door for the patenting of genetic material.[20] While products of nature are not eligible for patents, isolated DNA sequences are considered new compositions of matter and hence patent-eligible. One might argue that this liberalized patent policy is a by-product of the attitude of "technicity" identified by Heidegger.

The assignment of proprietary rights to a person's extracorporeal bodily material and DNA information would surely accelerate this expansion of commodfication and change people's perceptions about their intrinsic worth that should always transcend market forces and utilitarian "calculations." Commodification is demeaning to the body, which is integral to personhood. The body is not an aggregate of parts and genetic tissues extrinsic to the conscious self. The dualistic assumptions implicit in this viewpoint are incorrect. Rather, the person *is* a rational, sentient body. Bodily life along with its biological processes and components are in and of themselves personal values.[21] Hence the perception of the body (and its constituent parts) as a source of economic value to be bargained away for the best price implies a reduction of the person to just another piece of property that can be bought and sold. The person comes to see his or her body (and hence his or her personhood) as a means to a profitable end.

Accordingly, establishing a property interest in the body and its products has the potential to impair human dignity by diminishing respect for the intrinsic good of human life and bodily integrity. Such respect shapes key legal doctrines that resist any association of personhood and property. Property rights are imprudent also because they will induce the indigent to sacrifice their bodily integrity for small sums of money, perhaps to the detriment of their own health. To the extent that the creation of a new property right over one's "individual genome" intensifies the commodification of human life, it is not a welcome idea.

PATENTING THE GENOME

The other utilitarian argument against individual ownership is equally persuasive: its detrimental effects on research. The majority opinion in the *Moore* case affirmed this general sentiment by rejecting Moore's claim that his property right had been violated when doctors did not share the commercial gains they had obtained through the use of his surgically excised spleen cells. As we have seen, a key issue in this case was whether or not Moore owned his human tissue source along with the genetic information coded into his cells, but the court concluded that he did not have a valid ownership claim. The court's principal rationale was that the endowment of such a property right would hinder scientific research: "this exchange of scientific materials, which is still relatively free and efficient, will surely be compromised if each cell sample becomes the potential subject matter of a lawsuit."[22] In making its decision, the *Moore* court at least implicitly rejected the claim that researchers had a legal duty to share the bountiful rewards of that research with those like John Moore who contribute human tissue (or other samples).[23] Similarly, in *Greenberg v. Miami Children's Hospital Research Institute,* a federal district court found that the patent for the gene for Canavan's disease, which was discovered from Greenberg's tissue sample, did not violate the rights of the donor since that donor has "no cognizable interest in body tissue and genetic matter."[24]

While it may be appropriate to compensate Moore and other donors of valuable cell material, the status quo affirmed by these cases should remain in effect. Bestowing a new set of proprietary rights for tissues and the DNA segments that could be isolated from those tissues will inevitably yield a negative aggregate impact, which, though difficult to quantify, is likely to be quite substantial. This multitude of new rights, and the transaction costs necessary to enforce them, would create disincentives for research and development. As a result, these rights would put a stranglehold on basic "upstream" resources far removed from "downstream" diagnostic tests and therapies. Consider the retardation of research efforts that would be caused by the need for negotiations with multiple data owners, the payment of licensing fees to this web of owners, and the inevitability that some of them will act opportunistically. Thus, the barriers to biomedical research and probable impediments to patient testing for genetic disorders and therapies represent a heavy burden that cannot be justified in social welfare terms despite the benefits that may accrue to certain individual donors.

In addition, recognition of an intellectual property right ex ante in one's native DNA (before any manipulation of the cells to isolate a DNA sequence) represents a major but untenable paradigm shift. The donor's cells and tissue samples constitute the raw materials of genetic innovations, and intellectual property rights (such as a patent) should be granted only to those who take raw materials from nature and transform them into something that goes beyond what nature has provided. Individuals who donate tissues and cells from which DNA sequences are extracted do not engage in this type of creative research activity, which is both expensive and labor intensive.[25] A product should be eligible for a property right only if the prospective rights holder can demonstrate that this product "is not nature's handiwork, but his own."[26] It is difficult but not impossible to discern where nature ends and creative innovation begins, and this differentiation must remain a prime criterion for validating patent claims.

This position seems consistent with the Lockean justification for property rights, which maintains that authentic property rights are engendered by labor on unowned or common resources. Locke provides two reasons for his normative claim that a person's labor entitles that person to whatever is produced by means of that labor: (1) the right is derived from a prior property right in the labor produced by a person's mind and body; (2) a property right is deserved as a just return for the laborer's pains. According to Locke, when a person's labor is "annexed" to common resources, those resources are transformed and thereby removed from the commons, and the end result is a property right.[27] Assuming the legal status quo, the genome is equivalent to a commons, and the researchers' labor could be substantial enough to justify the assignment of a morally valid intellectual property right. In this case, the researchers connect their labor to this cellular and genetic raw material, which is transformed when a DNA strand is removed from the remainder of the strand and the cell's other parts. For this work of isolating the DNA segment, the laborer would appear to deserve a property right or at least warrant the presumption of such a right.

What does the law have to say about this? The legal landscape has shifted dramatically, thanks to the landmark case of *Association for Molecular Pathology v. U.S. Patent & Trademark Office*.[28] The patent holder in this case, a company called Myriad Genetics, claimed an isolated piece of DNA containing the nucleotide sequence that translates into either the BRCA1 or BRCA2 protein. Myriad had identified this DNA sequence that codes for these proteins and was awarded a patent. These patents for isolated DNA sequences have a "preemptive effect," since they exclude anyone from working with the BRCA genes without permission. Only Myriad can commercialize this discovery through the development of diagnostic screening tests, gene therapies, or other products. In addition to the patenting of this isolated DNA, the patent holder also claimed a product called cDNA, which is the "mirror image" of the DNA sequence. cDNA does not normally exist in the human body, and is naturally created only through the operation of certain retroviruses. Transforming normal DNA into cDNA, however, provides a more efficient tool for researchers and health care professionals who wish to study, diagnose, and treat the disease associated with a gene.

Along with its patents for the BRCA genes, Myriad was also awarded a patent for the method of determining whether a person is predisposed to the relevant form of cancer by comparing the person's gene sequence to the sequence in nature that codes for either BRCA1 or BRCA2. Finally, Myriad received a patent for the method of determining whether a particular cancer therapy is efficacious by growing cells containing the relevant gene and determining whether those cells grow more slowly when subjected to that therapy.

A lawsuit by a group of genetic researchers contested the validity of these patents, arguing that the BRCA genes are "natural human genes." They also claimed that the methods did not constitute patentable subject matter.[29] The plaintiffs maintained that the monopoly of these genes enabled by the patent interfered with the capability of patients to obtain better cancer-screening tests. The district court agreed with the plaintiffs and rejected all of the Myriad patents, reasoning that since the purification of natural DNA does not alter its inherent characteristics, isolated DNA remains a product of nature. The court concluded that "because the claimed isolated DNA [was] not markedly different from native DNA as it exists in nature, it constituted unpatentable subject matter."[30]

On appeal, the federal appeals court ruled in favor of the patent holder, reversing the decision of the lower court. The appeals court asserted that due to human intervention, the isolated DNA did exist in a distinctive chemical form and therefore was different from DNA in the human body (or native DNA). Since isolated DNA has this different chemical structure, it is eligible for a patent. Moreover, after decades of genetic patents and a pattern of firm judicial precedent including *Chakrabarty* and *Moore*, the court concluded that it could not now identify isolated DNA as nonpatentable and thereby disrupt the "settled expectations" of the scientific community.[31] As result, the Court declined to extend the "laws of nature" exception to include isolated DNA sequences. At the same time, the appeals court affirmed the two method patents, one for comparing and analyzing DNA sequences and the other for screening potential cancer therapeutics by way of changes in cell growth rates.

Arguably, a more balanced and legally nuanced outcome would have supported the patentability of the cDNA claims, and the patentability of the method claims, while holding that isolated DNA sequences or BRCA gene patents should be voided. Simply isolating the BRCA genes by extracting them from their natural location and incidentally changing their molecular structure in the process seems an insufficient basis for a patent. This innovation appears to still fall on the side of products of nature as something discovered but not invented. As an amicus curiae brief for the plaintiff stated, awarding a patent for the discovery of the BRCA gene is like awarding a patent for the discovery of a chemical element such as lithium. On the other hand, cDNA cannot be isolated from nature but must be created in the laboratory, so it should be patent-eligible. Similarly, assuming that the method claims meet the general criteria for method patents, they too should be considered patent-eligible so that researchers can reap the rewards from the application of its discoveries.[32] Such a solution balances the innovator's reward with the preservation of open access to the human genome for the sake of future research.

Hence, researchers or biotech companies like Myriad should not be able to obtain patent rights to isolated DNA sequences. Although different in molecular structure, those

DNA sequences have not been sufficiently modified, so they are still fundamentally the same entities that they were in their natural state. Patents will mean that other researchers are preempted from using these mutant genes for their own scientific work. As one dissenting judge in the *Molecular* case stated, "broad claims to genetic material present a significant obstacle to the next generation of innovation in genetic medicine—multiplex tests and whole genome sequencing."[33] The purpose of patent protection is to stimulate innovation, but sometimes too much protection can impede rather than promote innovation. Given these valid preemption concerns and the proximity of these modified genes to native DNA, the justification for BRCA patents is on tenuous moral and legal ground.

How does this conclusion square with a moral analysis that depends on the Lockean framework introduced earlier? Isn't a property claim legitimate, given the complexity of Myriad's discovery and the laborious efforts involved in the use of sequencing techniques to uncover the nucleotide order of the BRCA genes? Despite this effort there are two reasons why patents are not warranted according to Locke's framework. First, for Locke, labor gives rise to a property right only when it transforms and adapts something from the state of nature. There is some question in this case whether the labor of researchers like Myriad is transformative enough to warrant a property right, since it is hard to see that isolated DNA is markedly different from native DNA. Locke is always insistent that labor must put a "distinction" between what is worked upon and the commons, but does the discovery of a DNA segment that is unique to a group of individuals (such as the BRCA gene) create such a decisive distinction? If patents were awarded for these genes, why not for chemical elements like lithium, which also must be isolated for industrial applications but which "is the same element whether it is in the earth or isolated."[34]

Second, even if it could be argued that isolated DNA is distinct enough from native DNA, a patent would still be problematic from a Lockean perspective. While Locke believed in property rights based on labor, he did not support unlimited rights. Locke insists on an important condition limiting the acquisition of property, which is referred to as the sufficiency proviso. According to this principle, one can appropriate an object from the commons through labor only when "there is enough, and as good, left in common for others."[35] This proviso should apply to both physical and intellectual property rights. Moore frames this proviso in terms of weak Pareto superiority, which permits individuals to better themselves so long as no one is made worse off in the process. Since no one is harmed it is "unreasonable to object to a Pareto-superior move."[36] Thus, if the acquisition of an intangible work or patentable subject matter makes no one worse off in social welfare terms, compared to how they were before the acquisition, then an intellectual property right is valid. For most intangible works such as novels or poems, no one is made worse off by the acquisition (provided that the presumptive property right is given to the expression of ideas and not to the ideas themselves), and the labor creates a prima facie property claim to that work.

However, this is not the case with the patenting of isolated DNA sequences that cannot pass the Pareto superiority test. The patenting of the BRCA genes is not consonant with Locke's proviso because it does make others worse off by preempting them from using these valuable genetic resources. In this regard, it does not leave "enough and as good for others." When genes are patented, researchers are constrained from studying the genetic basis of a disease such as Canavan's disease without the payment of a steep licensing fee to the patent holder. In some cases, companies refuse to license their patents and products, and thereby foreclose research altogether.[37] Patents for diagnostic methods and therapies, on the other hand, have a much better chance of satisfying a Pareto-based proviso, if they are awarded properly and do not involve patenting laws of nature in ways that tie up the future use of those laws.

Broad gene patents like the ones awarded to Myriad do not satisfy the requirements of Locke's sufficiency proviso because they remove from the commons what others vitally need for genetic research. Hence, despite the arduous labor involved, the patents awarded in these cases lack a moral foundation since they do not enhance social welfare, and since others are made worse off through this acquisition and subsequent exclusion. The appeals court that decided the *Myriad* case may be right in asserting that it does not have a duty to rewrite the law, but patent policy should be adjusted to preclude this type of patent.[38]

CONCLUSION

Regardless of the valid privacy and social justice concerns, we must resist the impulse to recognize a property interest in a person's extracorporeal bodily material and its embedded DNA information. Preventing the misappropriation and misuse of this genetic data is a worthy objective, but this goal can be accomplished by implementing a more sophisticated informed consent mechanism coupled with legally mandated confidentiality requirements such as GINA. Because of the correlation of data that typically takes place with genetic information, privacy protection poses a formidable challenge, and promises of confidentiality may need some qualification. Since anonymity alone may not protect privacy, a search for more cogent methods should be undertaken. But proprietary control over our unique genetic identities should not be an option because of its detrimental impact on biomedical research.

Similarly, the proliferation of gene patents to biotech companies yields the same negative side effects, so they too cannot be warranted on social welfare grounds, despite the "settled expectations" of the scientific community. What is essential is a patent policy that preserves the intellectual commons rather than depletes it. Patenting the applications of one's discoveries such as diagnostic methods and therapies is morally suitable, because, unlike gene patents, these patents are compatible with the Lockean framework. A patent policy that permits patents for the application of genetic discoveries but not for the discoveries themselves would more properly balance the need for an innovator's just reward with the need for accessibility to common resources. The less patents interfere with research, the greater the opportunity to maximize the benefits from the Human Genome Project and promote the common good of human health.

NOTES

1. James Gleick, The Information: A History, a Theory, a Flood 11 (2011).
2. W. Klug & M. Cummings, Essentials of Genetics (2d ed. 1996).
3. *See* Ben Mitchell, *The New Genetics and the Dignity of Humankind*, *in* Human Dignity in the Biotech Century 60–74 (C. Colson & N. Cameron eds., 2004).
4. Carl Zimmer, *Hope, Hype, and Genetic Breakthroughs*, Wall St. J., Mar. 11, 2011, at C6.
5. C. Heeney et al., *Assessing the Privacy Risks of Data Sharing in Genomics*, 14 Public Health Genomics 17–25 (2010).
6. Mitchell, *supra* note 3, at 66–67.
7. Anton Vedder, *KDD, Individuality, Privacy, and Fairness*, *in* Readings in Cyberethics 462–70 (R.A. Spinello & H. Tavani eds., 2004).
8. Scholars like Annas have advocated such special protection. *See* G. Annas, *Privacy Rules for DNA Databanks*, *in* Genetics & Society 100–12 (P. Barker ed., 1995); *see also* E. Juengst, *Face Facts: Why Human Genetics Will Always Provoke Bioethics*, 32 J. L. Medicine & Ethics 76–98 (2004).

9. Onora O'Neill, Autonomy and Trust in Bioethics 101 (2002).
10. *Id.* at 105.
11. Herman Tavani, *Ethics at the Intersection of Computing and Genomics, in* Ethics, Computing, and Genomics 23–24 (H. Tavani ed., 2006).
12. This proprietary right might include a property right to one's extracorporeal cells along with an intellectual property right to the DNA information extracted from those cells.
13. Moore v. Regents of the Univ. of Ca., 793 P.2d 479, 510 (Cal. 1990) (Mosk, J. dissenting).
14. D. McLochlin, *Whose Genetic Information Is It Anyway?*, 19 J. Marshal J. Computer & Info. L. 609 (2001); *see also* Margaret Everett, *The Social Life of Genes, in* Information Ethics 226–50 (A. Moore ed., 2005).
15. A. Miller, *Personal Privacy in the Computer Age: The Challenge of New Technology in an Information-Oriented Society*, 67 Mich. L. Rev. 1203 (1969).
16. David Koepsell, Who Owns You: The Corporate Gold Rush to Patent Your Genes 167 (2009).
17. *Id.* at 168.
18. L. Sharp, *The Commodification of the Body and its Parts*, 29 Ann. Rev. Anthropology 287–328 (2001).
19. Martin Heidegger, Holzwege 237 (1950).
20. Diamond v. Chakrabarty, 447 U.S. 303 (1980).
21. For a persuasive defense of this viewpoint, see Patrick Lee & Robert George, Body-Self Dualism in Contemporary Ethics and Politics (2008).
22. *Moore*, 793 P.2d at 506.
23. *But see Donna Gitter, Ownership of Human Tissues: A Proposal for Federal Recognition of Human Research Participants' Property Rights in Their Biological Material*, 61 Wash. & Lee L. Rev. 257 (2004) (criticizing the *Moore* ruling and (like Mosk) advocating for the recognition of property rights for research participants).
24. Greenberg v. Miami Children's Hosp. Research Inst., 264 F. Supp. 2d 1064 (S.D. Fla. 2003).
25. *See* Robin Feldman, *Whose Body Is It Anyway? Human Cells and the Strange Effects of Property and Intellectual Property Law*, 63 Stan. L. Rev. 1377 (2011).
26. *Diamond*, 447 U.S. at 303.
27. John Locke, Second Treatise of Government § 27.
28. Ass'n for Molecular Pathology v. United States PTO, 653 F.3d 1329 (Fed. Cir. 2011).
29. Ass'n for Molecular Pathology v. United States PTO, 702 F. Supp. 2d 181 (S.D.N.Y. 2010).
30. *Id.* at 187.
31. *Ass'n for Molecular Pathology*, 653 F.3d at 1348.
32. Those criteria have been articulated in *Bilski v. Kappos*, 130 S. Ct. 1238 (2010).
33. *Id.* at 1368.
34. *Id.* at 1376.
35. Locke, *supra* note 28.
36. Adam Moore, *Intangible Property: Privacy, Power, and Information Control, in* Information Ethics 176–80 (1998).
37. Myriad itself has been accused of this behavior. *See* Christopher Holman, *The Impact of Human Gene Patents on Innovation and Access: A Survey of Human Gene Patent Litigation*, 76 UMKC L. Rev. 95 (2007).
38. As the majority asserts, "we are without competence to entertain arguments about the grave risks generated by genetic research." *See Ass'n for Molecular Pathology*, at 1394.

CHAPTER 8

Bioinformatics Collaborations: Key Considerations in Negotiating a Deal

Kimberly J. Wade

INTRODUCTION

Bioinformatics is increasingly a focus for pharmaceutical and biotechnology companies as these companies seek to more rapidly move products from discovery, to development, to the market. The volume of data that is available to scientists requires sophisticated software to analyze the data, as well as individuals with the expertise to mine this data. Many companies in the life science industry lack the technology needed to fully capitalize on this wealth of information. The time delay that in-house development would require places such companies at a competitive disadvantage. Thus, life science companies often look to information technology companies, which either have developed complex computer programs that fill the void or otherwise have the in-house expertise to develop such programs, to access and develop advanced bioinformatics capabilities.

Technology companies, from start-up companies to tech giants such as IBM, have recognized the opportunities that are available in the bioinformatics field, and have developed business models around providing services and software to pharmaceutical and biotechnology companies. While much of the activity in these fields revolves around fee-for-service arrangements or licensing of commercial software, information technology companies have also capitalized on opportunities to partner with pharmaceutical and biotechnology companies to develop customized technology that will assist in the identification and discovery of development candidates products.

The parties generally enter into a partnering relationship in the form of a collaborative development agreement or strategic alliance agreement where the parties agree to provide their respective expertise to develop a technology that can be used by the life science company in its drug discovery and development programs. For example, in October 2011 Bayer HealthCare Pharmaceuticals and MediSapiens Ltd. announced that the parties had reached an agreement to develop a platform for the analysis of cancer genomics data.[1] In the press release issued by MediSapiens for this agreement, MediSapiens CEO Tommi Pisto noted that "[g]enomics data is increasingly used both in preclinical and clinical research, but lacks software tools that can be adapted to various research environments." He went on to note that the MediSapiens platform and know-how in designing software for bioinformatics would be tailored to meet the needs of oncology research at Bayer. These types of collaborations are increasingly more common for life science companies.

WHERE DO WE START?

Once a company has decided to seek a collaboration partner, whether that company is an information technology company or a pharmaceutical company, there are many preparatory steps that the company should take to ensure that it identifies the type of collaboration that it desires and the type of partner with which it will work best. Before approaching potential partners, a company should identify its business objectives, understand the relevant market, and research potential partners.

The business objectives of a company will vary depending on which side of the partner equation it falls. However, one basic question for both the information technology company and the life science company is how this collaboration will benefit the company. Put another way, what must the company gain—financially, in technical expertise, or otherwise—in order to make this partnership worthwhile?

An information technology company may have an objective to establish a higher profile in the bioinformatics field and to gain access to data in order to develop or fine-tune its software. A partnership with a large pharmaceutical company could serve both purposes by validating the information technology company as a viable bioinformatics partner and by giving the company access to scientists who can help in the technical development of the software. In some cases, such partnerships are necessary in order to obtain funds to keep the company solvent. For a pharmaceutical company, the ultimate benefit is to identify new resources that will allow it to identify, develop, and commercialize products.

Knowledge of the relevant market and terms for partnering agreements can go a long way in facilitating the negotiation of a collaboration agreement. Before starting the negotiation process, a company should be aware of standard terms in collaboration agreements and current practices and trends in the industry. Much of this knowledge may be present within the institution; however, to the extent that a company does not possess such institutional knowledge, there are many resources to which a company may avail itself. A company should identify recent agreements, including publicly filed agreements, which are similar to the transaction that it intends to negotiate in order to familiarize itself with the terms. Additionally, trade journals can provide guidance as to recent trends in the industry, which may prove useful in negotiations.

It is difficult to find the ideal partner. Unlike a simple buy/sell transaction where issues revolve primarily around pricing, the evaluation of a collaboration partner requires the weighing of many more factors. Ideally, both parties will bring to the table different technology and capabilities that offer value to the other party. Another consideration in selecting a partner is the culture of the two companies and how the companies will work together. It is important to also conduct research on a potential partner, utilizing publicly available information. Reviewing a company's public filings, technology-related press releases, website, and other resources may provide useful insight into the history and direction of the company.

Once a company has identified a partner and come to a basic agreement on negotiating a transaction, it is time to commit the key terms to paper in the form of a preliminary agreement. A preliminary agreement may take the form of a term sheet, memorandum of understanding, letter of intent, or similar agreement. These preliminary agreements are typically nonbinding, although care should be given to specifically state whether the parties intend for the terms to be binding or nonbinding. A preliminary agreement allows the parties to hash out some of the key terms to be included in a definitive agreement before committing to the full negotiation of such agreement.

KEY PROVISIONS IN COLLABORATION AGREEMENTS

Many of the key provisions found in collaboration agreements between a life science company and information technology company are found in discovery-stage collaboration and licensing deals in the life science industry. In addition, where the life science company receives a license for the software, these collaboration agreements also include certain customary terms found in software license agreements. Some of the more common key provisions found in bioinformatics collaboration agreements are identified below.

Exclusivity

An important initial question for both the licensor and licensee is whether the licensee will be granted exclusive or nonexclusive rights under the technology. Exclusivity versus nonexclusivity is an important issue where the licensor is licensing a platform technology or software that can be used broadly by many companies.

In a typical life science collaboration agreement, the licensee will obtain an exclusive license under the licensed technology in order to develop licensed products. This is important to the licensee in order to ensure that the technology is not licensed to its competitors. In a typical information technology agreement, however, the licensor grants a nonexclusive license to use the technology. There is therefore somewhat of a gap to bridge in negotiating a bioinformatics collaboration agreement.

One way to resolve this issue is for the licensor to bifurcate the license grants with respect to different types of technology, giving the licensee an exclusive license with respect to the results of the collaboration but a nonexclusive license with respect to the licensor's existing technology. Depending on the type of technology licensed and the stage of the licensee's development activities, the parties may also come to an agreement on a field-restricted exclusive license that gives the licensee exclusivity with respect to specific targets or products and/or in a specific field of use, but that does not otherwise encumber the licensor's technology generally. The licensor's ability to grant any exclusive rights will, however, be subject to any other rights it has granted to third parties prior to such grant.

Another aspect of exclusivity found in many agreements is a noncompete covenant given by one or both parties that it will not work, directly or indirectly, to research, develop, or commercialize targets or products that are or could be competitive with the targets or products that are the focus of development under the collaboration agreement. A noncompete provision in the collaboration agreement can minimize future conflict between the parties by ensuring that the parties are focused on the purpose of the collaboration and not distracted by competing opportunities. These provisions are often highly negotiated and may extend for the full term of the agreement or some subset of time that still allows the products being developed under the agreement to have a "head start" on other products that either party may develop after the noncompetition term has ended.

Field

As the name implies, a field of use provision in a collaboration agreement limits the licensee's rights in the licensed technology to a particular field, use, or application. The "field" generally is defined in the definitions section and then used to limit the scope of the licensed rights where applicable throughout the agreement. The scope of the field is very important to both parties in the collaboration. A licensee will seek the broadest field definition so as to enable it to exploit the licensed rights as broadly as possible. Conversely, the licensor tends to seek a narrow definition of the field. However, if the field definition is drafted too narrowly, the licensee may lack sufficient rights to fully exploit the technology. This tension arises primarily in exclusive licenses where the licensee is interested in using the technology only in a particular area, but the technology can be applied in a wide variety of fields.

To the extent that the licensee is granted an exclusive license, agreeing to a field definition that is too broad could cause the technology to sit dormant in fields in which the licensee has no intention to maximize the technology that could otherwise prove to be lucrative to the licensor. For example, a licensee may be granted rights to use the technology in all fields of use, but it may plan to use the technology only to identify and discover targets or products for therapeutic uses in oncology. The licensor in this case should seek a narrower field definition so as to give the licensee the rights it needs to exploit the technology in its intended area of use, while the licensor retains the rights to exploit the technology for other uses. In the exclusive license context, to assuage the licensee's concerns on potentially missing out on the excluded fields, the licensor can grant the licensee an option to expand the field or a right of first refusal or negotiation if the licensor proposes to license rights to a third party outside the licensed field. Negotiating the definition of the field requires communication between the parties so that each party understands the needs and objectives of the other party.

Payment Terms

The economic terms of a collaboration agreement are often negotiated by the parties early on in the term sheet stage of discussions. There is no clear right or wrong way to structure payments under the agreement, and the structure in large part will depend on the bargaining power of the parties. Payment can take many different shapes in a collaboration agreement. Some payment types typically found are up-front or signing fees, research and development funding, milestone payments, maintenance fees, royalties, and reimbursement for patent prosecution and maintenance expenses. In addition, the parties may agree to other less common financial terms such as lines of credit or equity investments. Whether an agreement includes all or a subset of these types of payments, and the amounts of these payments, is a heavily negotiated point.

An up-front or signing fee is fairly standard and acts to compensate a licensor for its investment in the technology to date, in addition to the grant of rights under the agreement. In most cases, the licensor wants to receive as much of the financial consideration as possible as an up-front signing fee. In contrast, the licensee generally prefers to defer compensation by paying smaller amounts up front and allocating larger amounts to later milestone or royalty payments after the technology has been proven and/or viable products have been identified. This is particularly the case in collaboration agreements where the parties will develop new technology or improvements to existing technology. Often, a licensor will be eligible to receive a greater cash return if it accepts a smaller up-front payment and receives more in milestones and royalties later in the life of the agreement. The risk, of course, is that the development work under the agreement will not be successful or that an early termination of the agreement would deprive the licensor of such later payments.

Milestone payments are very common in collaboration agreements. Milestones may be based on activities of one or both of the parties under the agreement, including achievement of specified goals of the development under the agreement or with respect to the development and/or commercialization of products resulting from the activities under the agreement. The parties should be clear as to what constitutes achievement of a particular milestone and reporting requirements upon achievement of a milestone.

Royalties are another common component of collaboration agreements. Royalties are typically based on net sales of products that are developed as a result of the use of the technology licensed under the agreement. The royalty rate imposed on discovery-stage collaborations will typically be a low rate, which increases for later stages of development. The parties may also include deductions against the royalties and/or milestones payable under the agreement for payments that the licensee is required to pay to third parties for the grant of rights needed to develop or commercialize the product, and due to generic competition.

Payment terms may also include a reimbursement for research-and-development costs incurred under the agreement. These costs may include a party's internal costs as well as out-of-pocket costs paid to third parties. The parties can share research-and-development costs in a few ways. Often, where a larger company is working with a smaller company, the larger company will agree to reimburse the smaller company for its research-and-development costs incurred in performing work under the agreement. These costs generally are required to be incurred in accordance with an agreed-upon budget for such activities. In other cases, the parties may agree to share the research-and-development costs equally, or each party will be responsible for the costs that it incurs under the agreement.

Many licensors find annual (or other periodic) fees to be a good way to encourage the licensee to utilize the licensed technology. The basic idea behind this is that a licensee would be willing to sit on a technology if there were no ongoing payment obligations related to the license, but if the licensee is required to pay a periodic license or maintenance fee, it is much more likely to either use the technology or terminate the agreement. However, in cases where a licensee has exclusive rights, it may be willing to pay such fees in order to prevent a competitor from gaining access to the licensed technology.

Record Keeping and Audit Rights

In bioinformatics collaboration agreements where the financial terms of the transaction include milestone payments for the development of a product or royalties based on net sales of a product, it is important for the licensee to have an obligation to maintain records that identify all products that are developed as a result of using the licensed technology. The licensee may be sensitive to the level of detail of the records that it is required to maintain so that it does not create on onerous obligation outside of its normal record-keeping practices. These records are, however, important to the licensor to help establish its rights to receive compensation based on the development or commercialization of products arising from the collaboration.

Hand in hand with the licensee's obligation to maintain records is the licensor's right to audit these records so that it can identify all milestone or royalty-bearing products. The licensee will generally insist on limitations on the audit rights. Some of the more customary limitations include limiting the right to audit the records to no more than once a calendar year, limiting the number of days for the audit to take place, and requiring that the audit be subject to confidentiality obligations. Additionally, a licensee may have concerns that the licensor may in the future be acquired by one of the licensee's competitors, and this audit right would give such competitor access to competitive information. This can be addressed by having the licensee exercise its rights through an independent auditor that is reasonably acceptable to both parties.

Intellectual Property

A key point of negotiation in any partnering agreement is which party will own the intellectual property that is developed under the activities of the agreement. Many permutations of ownership can be utilized in the partnering agreement. The following are some of the most common allocations of ownerships.

Rights Based on Inventorship

This is possibly the most straightforward allocation of rights. In this scenario, ownership of intellectual property rights or technology would be based solely on inventorship or authorship. Neither party would acquire ownership rights in intellectual property or technology invented or authored solely by the other party.

Rights Based on Background Technology

Under this scenario, each party retains ownership to all of its background intellectual property and technology (i.e., intellectual property and technology that it developed or acquired prior to entering into the agreement or during the term of the agreement if developed or acquired outside the scope of the agreement). In addition, such party owns all intellectual property and technology that constitutes an improvement to its background intellectual property and technology.

Rights Based on the Specific Field

Under this scenario, ownership rights to intellectual property and technology are determined based on the specific field. One example in a collaboration agreement between a life science company and an information technology company would be that the information technology company would own all intellectual property and technology developed under the agreement that relates to software and algorithms, while the life science company would own all intellectual property and technology related to targets, product candidates, and products.

Joint Ownership

In this scenario, each party would own a 50 percent interest in any intellectual property or technology developed under the agreement. Joint ownership requires the parties to address issues that arise under intellectual property laws with respect to the rights of joint owners. For example, the patent laws in some countries require joint owners to account to each other for the exploitation of jointly owned patents and/or they require that in order to enforce the joint patent against infringers, both owners must be joined to the suit. Also, the type of technology developed may also be covered under copyright laws, which may establish different rules for joint owners than the rules under applicable patent laws. These issues may be surmounted, but they require careful consideration by the parties to the agreement.

Diligence

Diligence provisions are often included in agreements to ensure that the parties carefully advance the agreement objectives. The diligence provisions can act as a safety net of sorts to ensure that the parties get the benefit of their bargain. Diligence obligations are usually based on a "diligent efforts" or "commercially reasonable efforts" standard, which is defined in the agreement. The definition may be either an objective standard, based on what another company in the relevant industry would do to achieve an objective, or it may be a subjective standard, based on what a party to the agreement normally does to achieve an objective. Additionally, the parties may agree to specific timetables for the completion of activities under the agreement.

Governance

Most collaboration agreements will have governance provisions establishing joint committees, comprised of representatives of both parties, to oversee the collaborative aspects of the relationship. There may be multiple levels of committees or one single governing committee. One of the more commonly found committees is a joint research-and-development committee. This committee is tasked with managing the parties' activities at the science or engineering level. Sitting above the joint research-and-development committee is a joint steering committee, which typically is tasked with overseeing the overall collaboration and resolving any conflicts that arise between the parties. The parties may form other committees as appropriate for the collaboration.

Governance provisions are important to both parties in a partnering relationship, as they provide not only a mechanism to work together, but also for the parties to protect their economic interests and maximize their rights under the agreement. The licensee wants to protect its investment in the technology and have the freedom to identify, develop, and commercialize any resulting products. The licensor wants to ensure that it receives all consideration under the agreement as timely as possible, and that it does not negatively compromise the licensed technology.

One of the more heavily negotiated points related to the governance provisions is how decisions of the committees will be made. In general, the committees act by consensus, but when the committee members cannot agree on an issue, how will the parties move forward? One option is to have the parties deadlocked until an agreement can be reached. This forces the parties to come to some compromise in order to move forward. Where a party invests significant resources in the collaboration, this outcome may be unacceptable and may require that the agreement specify that such party have the deciding vote on either all issues or specific issues that arise under the agreement. Other dispute resolution mechanisms also may be workable under the governance structure, such as the appointment of an independent expert to decide disputes of a technical nature, or the submission of a dispute to baseball-style arbitration (i.e., where both parties submit their proposed resolution of the dispute to an arbitrator, and the arbitrator chooses which party's proposal will win the day).

Term and Termination

When negotiating an agreement, neither party expects the collaboration to fail, but careful attention should be paid to the rights of each party to terminate the agreement and the consequences of any such termination.

Standard termination rights include termination for an uncured material breach, termination in the event of a bankruptcy of one of the parties, and termination for convenience. The first two termination rights are event-based and fairly straightforward. The right to terminate the agreement for convenience is usually a right granted to the licensee, but there may be limitations on this right. For example, for some initial period of the agreement, the licensor may not want the licensee to have the right to terminate the agreement. Additionally, a licensor may negotiate for a termination fee if the licensee terminates the agreement for convenience.

To the extent that the licensor is entitled to milestone or royalty payments based on products developed using the licensed technology, the licensor should pay particular attention to the rights of each of the parties with respect to those products following termination. Once the licensed technology has been utilized to identify or develop a product or product candidate, the licensee may not have a further need for a license under the technology. The licensee could therefore elect to terminate the agreement, as well as all future financial obligations to the licensor, and yet continue to develop and commercialize products that otherwise would have been subject to such financial obligations. The licensor should consider including provisions in the agreement that (1) require the licensee to identify all milestone/royalty-bearing products upon termination of the agreement; (2) restrict the licensee's ability to use the licensed technology going forward; and (3) restrict the licensee's ability to further develop or commercialize such products. The parties should also pay close attention to which rights survive termination. For example, if the licensor has been granted a license on improvements that the licensee made to the licensed technology, the licensor will want this license to survive. Other standard rights that survive termination include confidentiality and indemnification obligations. It may also be appropriate, subject to applicable anti-competition laws, to have any noncompete or nonsolicit covenant survive termination.

CONCLUSION

A strategic alliance between a life science company and an information technology company can be mutually beneficial to both parties. This chapter outlined some of the key provisions found in collaboration agreements for these types of alliances. Although the provisions included in such agreements are fairly standard, the specific language and terms are subject to the negotiations of the parties and therefore are extremely variable. The collaboration agreement should be negotiated in a manner to ensure that the parties' joint intent is properly reflected and the parties' interests are protected.

NOTE

1. Press Release, MediSapiens Ltd., MediSapiens and Bayer HealthCare to Develop Platform for Managing and Analyzing Genomics Data (Oct. 11, 2011), http://medisapiens.com/press-releases/medisapiens-and-bayer-healthcare-develop-platform-/.

CHAPTER 9

University Research and Licensing

Jennifer Carter-Johnson, Jeffrey S. Carter-Johnson, and Jorge L. Contreras

INTRODUCTION

Research universities have traditionally been catalysts for technological innovation, particularly in new and emerging industries. A recent report on the management of university intellectual property confirms this historical role, stating that universities "have a lengthy track record of providing dynamic environments for generating new ideas and spurring innovation, and for moving advances in knowledge and technology into the commercial stream where they can be put to work for the public good."[1] Products ranging from the Gatorade® sports drink to the polymerase chain reaction (PCR) gene sequencing technology have emerged from university laboratories. University-based research played a major role in the growth of the early biotechnology industry and has made notable contributions to industries such as computer software, medical devices, and the Internet.[2] In the United States, universities and other research institutions spent almost $54 billion on research in 2009[3] and, of the top 50 holders of U.S. patents in the "biotech and pharma" field, in 2011 eight were U.S. universities and seven more were U.S. and non-U.S. governmental or quasi-governmental research institutions.[4] In 2009, universities accounted for over 35 percent of research in the United States and 53 percent of basic science research.[5]

Against this backdrop, it is not surprising that much of the research relating to bioinformatics is being performed at research universities. Universities are natural incubators for bioinformatics research, as this research requires combinations of computational and biological expertise. Bringing together these two disparate technological backgrounds, numerous universities in the United States and internationally have established interdisciplinary programs of study and research in bioinformatics.[6] This torrent of university bioinformatics research produces large amounts of intellectual property that universities can use, often licensing the technology to commercial enterprises.

In this chapter, we first summarize several modes of university technology development and licensing. Next, we describe the evolution of university technology commercialization and the Bayh-Dole Act of 1980, which is widely credited with establishing the intellectual property structure of current university licensing and technology transfer. We then discuss some important legal and intellectual property considerations relevant to the development, commercialization, and licensing of university technology. While this treatment is necessarily brief, we hope that it may serve as a useful tool both to those who are considering collaborating with, or licensing technology from, a research university, and to university researchers who are contemplating the path to commercializing their bioinformatics innovations.

MODES OF UNIVERSITY RESEARCH AND TECHNOLOGY TRANSFER

University-based research in bioinformatics takes place in a variety of funding and collaboration structures. The particular structure governing a research project will have a significant impact on the intellectual property rights and technology transfer procedures applicable to that project. In this section, we outline several common modes of university research funding and technology transfer that are prevalent in the United States today. Funding organizations should be aware of the norms and structures of university research as described in this and subsequent sections when deciding whether funding university research will adequately promote their policy and intellectual property goals. Additionally, those interested in licensing university-owned intellectual property should also be aware of the norms and structures of university research as these will affect the licensing terms under which the licensor can utilize the technology.

Grant Funding

The U.S. federal government funds between 62 percent and 68 percent of university research in the United States, primarily through grant mechanisms.[7] Federal grants are typically awarded and administered by executive branch agencies such as the National Institutes of Health (NIH), Department of Energy (DOE), Department of Defense (DOD), National Air and Space Administration (NASA), National Science Foundation (NSF), and Department of Agriculture (USDA). For instance, between 2004 and 2010, programs within the NIH have awarded funding of more than $500 million to research related to bioinformatics.[8] While this funding is not directed exclusively to universities, university researchers are the beneficiaries of substantial grant funding relating to bioinformatics.

Industry Sponsorship

In addition to federal grant funding, a significant portion of university research is supported by private industry. In 2009, private industry provided $4 billion in funding for academic scientific research.[9] Such support can take two basic forms: general support and sponsored research. General support consists of unrestricted or earmarked contributions by industry to particular universities or research programs. Under such a model, the corporate donor, while likely obtaining public relations and other intangible benefits, typically does not gain the right to direct or commercially exploit the results of the university's research. Sponsored research, on the other hand, is more akin to a contracted research arrangement between the university and the corporate sponsor. The sponsor funds the university's conduct of a specific research project, sometimes in collaboration with the sponsor's own scientists, and typically obtains the right, or an option to license the right, to commercialize the resulting technology. Sponsored research arrangements are not uncommon, particularly in the life sciences. One study found that in 2000, 28 percent of university faculty in the life sciences received funding from private sponsors.[10] These arrangements, however, must be structured carefully to avoid disputes regarding inventorship and ownership of discoveries. In one recent case, Vanderbilt University scientists were held not to be coinventors on a patent covering the blockbuster drug Cialis, which they helped to develop under a sponsored research agreement with Glaxo.[11]

Licensing and Technology Transfer

Though significant research activity is undertaken at universities, their educational and research missions do not typically permit them to engage actively in commercial activity. Thus, in order to put university research to commercial use, universities must license or transfer technology to the private sector. To do this, most universities have established technology licensing offices (TLOs) responsible for evaluating the commercial promise of each new university invention, making decisions regarding patenting, identifying appropriate commercial partners, negotiating suitable license and option agreements, and then distributing the resulting royalties and other economic gains within the university.[12] After deducting the TLO's overhead, patenting costs, and the like, most universities allocate royalties in varying percentages among the responsible inventors, their academic departments, and the university at large.

In recent years, TLOs have displayed considerable activity. Data from a survey conducted by the Association of University Technology Managers (AUTM) indicates that, in fiscal year 2009, 4,624 licenses were executed by responding university TLOs.[13] During this period, 12,109 new U.S. patent applications were filed by responding TLOs.[14] Evidence indicates that 50 percent to 75 percent of TLO patenting and licensing activity falls within the biosciences and pharmaceutical fields, as opposed to fields such as software and electronics which account for less than 10 percent each.[15] It is unclear how bioinformatics technologies may have been classified in this and similar surveys since bioinformatics technologies straddle these two spheres.

While most universities, including research powerhouses such as Stanford, MIT, and Harvard, operate their TLOs as internal groups, sometimes falling under the jurisdiction of the university counsel or the office of the provost and sometimes operating semi-autonomously, others have elected to establish independent entities to manage intellectual property emerging from university labs.[16] The most notable of these is the University of Wisconsin–Madison, whose Wisconsin Alumni Research Foundation (WARF) was established in 1925.[17] WARF granted its first commercial license to the Quaker Oats Company in 1928 for a Vitamin D supplement developed to combat the childhood disease rickets.[18] Today, WARF enters into approximately 100 commercial licensing agreements per year and has contributed nearly $1 billion in net revenue to the university.[19]

Spin-Outs

In many cases, the most likely industrial licensee of a university invention is an established enterprise actively pursuing the development of products in the relevant field. Sometimes, however, established industrial partners may not exist, particularly when technologies are in new and emerging fields. In these cases, university researchers, working with external advisors and funders, may form start-up companies to commercialize the discoveries generated by their labs. According to AUTM survey data, 596 start-up companies were formed based on university-owned intellectual property in 2009, up from 241 start-up companies in 1994.[20] These companies are referred to as university "spin-outs," and AUTM reports that in 2008, more than 16 percent of university technology licenses were granted to such start-up companies.[21] In addition to licenses of university intellectual property, such spin-outs often make use of university-owned facilities and equipment, as well as the services of academics, technicians, and graduate students. University spin-outs have attracted significant public attention in recent years, both due to the phenomenal success of a handful of such ventures[22] and the potential conflicts of interest that plague academic investigators who actively participate in corporate research.[23]

Patent Pools and Commons

When multiple entities each hold patents necessary to exploit a single technology, a situation referred to as patent "stacking," or a patent "thicket" or anti-commons, may be said to exist.[24] In order for a producer to implement the technology in a product, it must obtain licenses from multiple parties, each acting independently and each seeking to maximize its gains. The sum of these individual demands may be excessive in relation to the overall value of the product being produced and thus may impede commercialization.

In order to address patent-stacking concerns, patent pools and commons have been developed. A patent commons occurs when a group of patent holders voluntarily commit not to assert patents relevant to a specific field, subject to certain conditions, thus allowing members of the commons and the public to avoid patent-stacking issues for products. A close relative of the patent commons is the patent pool, in which groups of patent holders may aggregate their essential patents into patent pools, which are then licensed and administered on a collective basis. Well-known patent pools exist in the areas of consumer electronics and digital Columbia University is one of the original patent holders in the large pool responsible for licensing use of the ubiquitous MP3 data compression standard. The formation of patent pools is complex and involves the application of antitrust analysis well beyond the scope of this chapter.[25]

UNIVERSITY RESEARCH AND THE BAYH-DOLE ACT

Due to the dominance of federal funding of university research, inventors and investors interested in bioinformatics research and technologies must understand the regulations surrounding the dissemination of federally funded research. In this section, we discuss specific practical, legal, and intellectual property considerations that arise in the context of the federal funding described above under the heading "Grant Funding."

A Brief History

Until World War II, university research in the United States tended toward the theoretical and received relatively modest governmental support.[26] With the advent of the Manhattan Project, however, federal funding for research, and applied research in particular, increased dramatically.[27] In the decades that followed, numerous federal agencies began to fund university research; today, the majority of university research is funded by the federal government, which contributed more than $32 billion to the research budgets of universities and nonprofit research institutions in 2008.[28]

Prior to 1980, intellectual property rights in federally funded inventions were governed by the rules of individual funding agencies and often inured to the agencies themselves.[29] Yet the federal government rarely put these inventions to commercial use; it is estimated that of the 30,000 federally owned patents in existence prior to 1980, only 5 percent were ever licensed to industry and even fewer used in commercial products or services.[30] In response to this perceived underutilization of federally funded research, the Bayh-Dole Act[31] was enacted in 1980. The purpose of the Act was to provide a consistent patent policy in regards to federally funded research and to promote the commercialization of resultant technologies.[32] The Act effected a major change in U.S. policy by allowing universities, small businesses, and other research institutions to retain ownership of inventions resulting from federally funded research. In exchange for this grant of ownership, the Act requires these entities to apply for patent protection in the United States and abroad, and imposes penalties for failing to take effective steps to achieve "practical application" of the inventions.[33]

Requirements of the Bayh-Dole Act

In exchange for giving universities the right to retain ownership of their federally funded inventions, the Bayh-Dole Act imposes a number of obligations. Given the pervasiveness and magnitude of federal research funding in the United States, most universities have incorporated the requirements of the Act into their standard technology development and licensing practices. The principal among these are described below.

Invention Disclosure

The Act and its implementing regulations require that each federally funded institution disclose to the relevant funding agency each invention reduced to practice within two months after it becomes known to the institution's patent administration personnel.[34] In order to support this obligation, each institution is also required to implement written agreements with its technical personnel (including faculty, technicians, and students) requiring them to disclose all such inventions to the TLO.[35] Typically such agreements, which may be implemented in signed contracts or binding policy documents, also include an explicit assignment of intellectual property rights from the inventor to the university or a requirement that the inventor so assign upon request by the university.[36]

Each university TLO submits invention disclosures to the applicable funding agencies, typically through the federal government's iEdison interagency web-based system, which accepts submissions for 18 different federal agencies.[37] Invention disclosures and other information submitted to a federal agency pursuant to the Bayh-Dole Act are treated as privileged and confidential and are not disclosed outside of the agency.[38]

A university's failure to comply with the disclosure requirements of the Act can result in the government's receiving title to the relevant invention.[39] In at least two litigated cases, courts have prohibited institutions from enforcing patents following a failure to comply with the disclosure requirements of the Bayh-Dole Act on the basis that the plaintiffs never acquired title to the patents in suit.[40] However, even in cases in which the government receives title to a federally funded invention, the university retains a nonexclusive, royalty-free, worldwide license to exploit such invention.[41]

Patent Election

A university may elect to retain title to any invention disclosed to the federal government within two years of making such disclosure.[42] If the university elects to retain title, it must file a patent application covering that invention in the United States prior to the expiration of any statutory bar date, and in any other countries in which it elects to retain title.[43] If it fails to make such filings, the government may receive title to the relevant invention.[44]

This is not to say, however, that universities file patent applications covering every invention that is disclosed by their researchers. In fact, according to AUTM, over 20,000 invention disclosures were filed across all responding U.S. research universities in 2009, whereas fewer than 10,200 new U.S. patent applications (both provisional and nonprovisional) were filed in the same year.[45] In many cases, the potential commercial value of an invention may be small, and the university's educational and research missions may better be achieved by permitting the researcher to publish the relevant findings and/or to release the invention, for example, on an "open source" basis. If a university wishes to discontinue prosecuting a patent application or maintaining a patent that was developed using federal funding, it must so notify the federal agency.[46] While such a notification technically gives the government the right to receive ownership of the invention, in practice governmental agencies rarely exercise this right.

A related issue concerns a university's right to an invention under the Bayh-Dole Act when an investigator purports to assign the rights in that invention to a commercial research sponsor. This issue recently arose when a Stanford University researcher, Mark Holodniy, entered into a sponsored research agreement with Cetus Corporation (now part of Roche).[47] Under the agreement, the researcher assigned his rights in an invention pertaining to an AIDS therapy to Cetus, in violation of Stanford's intellectual property policy and Holodniy's previous agreement with Stanford that he would assign rights to Stanford after any invention. Holodniy developed a technique for measuring human immunodeficiency virus (HIV) load in the blood of patients while he was collaborating with Cetus employees, and Stanford went on to obtain patents for the technology.[48]

Stanford subsequently sued Roche Molecular Systems, who had acquired Cetus's assets relating to HIV research, including any rights in Holodniy's HIV research with Cetus, for infringement of Stanford's patent.[49] Stanford argued that the patent ownership provisions of the Bayh-Dole Act preempt private assignments such as that effected by Dr. Holodniy to Cetus, or in the alternative, that Stanford's agreement with Holodniy that he assign his rights acted as an assignment.[50] The Supreme Court found that the Bayh-Dole Act did not result in automatic vesting of the intellectual property to the university.[51] Instead, the property rights vest initially in the inventor and therefore a valid assignment is required for the university to own the intellectual property.[52] Additionally, in affirming the Court of Appeals judgment, the Supreme Court distinguished an obligation to assign from an actual, valid assignment.[53]

Government Rights

In addition to the right to receive ownership of inventions as described above, the federal government retains several additional rights in federally funded inventions. First, it retains a nonexclusive, paid-up license to practice, or have practiced, any such invention for or on behalf of the United States anywhere in the world.[54] Second, the government may exercise so-called march-in rights, under which it may compel a university to license an invention to one or more third parties if necessary to alleviate health or safety needs, if the university has not taken effective steps toward the commercialization of the invention, or if the U.S. manufacturing requirements described below are violated.[55]

In practice, the federal government has never exercised its march-in rights under the Act, though there have been several instances in which third parties have petitioned federal granting agencies to exercise those rights. The first instance occurred in 1997 when CellPro, Inc., petitioned the NIH to exercise march-in rights against Johns Hopkins University.[56]

CellPro's goal was to obtain a license to four patents that Johns Hopkins had previously licensed exclusively to Baxter Healthcare. The NIH determined that the exercise of march-in rights was not warranted because Baxter Healthcare had used reasonable efforts to make a product manufactured under the patents available.[57] In 2004, two individuals petitioned the NIH to exercise march-in rights against Abbott Laboratories, which had received NIH funding to develop the AIDS drug Norvir, after Abbott increased the retail price of the drug by approximately 400 percent.[58] Again, the NIH determined that the patentee had used the requisite efforts to achieve practical application of the federally funded invention, and it further commented that the exercise of march-in rights "is not an appropriate means of controlling prices."[59]

Another request for the exercise of march-in rights was made to the Department of Health and Human Services with respect to the drug Fabrazyme, which is used to treat the rare disorder Fabry's disease. Fabrazyme was created in part with NIH grant funding and is currently the only FDA-approved treatment for Fabry's disease.[60] The manufacturer of Fabrazyme, Genzyme, was forced to shut down its primary Fabrazyme manufacturing line due to contamination, which resulted in shortages of the drug and rationing to patients since June of 2009.[61] Fabry's disease patients subsequently filed a petition with the Department of Health and Human Services asking the federal government to exercise its march-in rights to allow an alternative manufacturer to produce the compound.[62] NIH denied the petition, however, reasoning that any alternative manufacturer would face substantial and time-consuming regulatory hurdles that would not soon result in an increased supply of the drug.[63]

U.S. Manufacturing

The Bayh-Dole Act prohibits the owner of an invention made using federal funding from exclusively licensing the use of that invention in the United States unless the licensee agrees that all products embodying the invention, or produced through use of the invention, will be manufactured substantially in the United States.[64] This provision is essentially a "Buy American" initiative intended to promote U.S. industry and has been criticized as outdated in today's global economy.[65] The U.S. manufacturing requirement may be waived by the funding agency if domestic manufacture is not "commercially feasible" or if efforts to identify U.S. manufacturers have been unsuccessful.

Nonassignment

The Act expressly prohibits universities from assigning rights, i.e., transferring actual ownership of the patent, in federally funded inventions to third parties without the approval of the funding agency.[66] An exception is made only for assignments to patent management entities such as the University of Wisconsin–Madison's WARF. This restriction often causes confusion among inexperienced venture capitalists and angel investors who argue, sometimes vociferously, that university spin-out companies should obtain full ownership, rather than a mere license, of the fundamental patents underlying their business. This perception is also widely shared by non-U.S. investors, who are accustomed to dealing with non-U.S. university spin-outs, which are typically not subject to nonassignment prohibitions under local legislation.

Royalty Sharing

The division of economic returns from university technology is typically handled internally by the university through its TLO. The Bayh-Dole Act requires only that universities share royalties with individual inventors, without specifying the level or form of such sharing, and that the balance of these proceeds (after payment of expenses) "be utilized for the support of scientific research or education."[67] Royalty-sharing arrangements vary widely among institutions. For example, Stanford University allocates the first 15 percent of net license revenue (after patenting costs) to its TLO, then splits the remaining 85 percent in three equal parts among the inventors (in equal shares), their departments, and the university; Washington University in St. Louis allocates 25 percent to its TLO, 35 percent to the inventors, and 40 percent to the university; and Rice University allocates 37.5 percent to the inventors, 14 percent to their departments, 18.5 percent to the graduate education function, and 30 percent to the university.[68]

While these arrangements are typically invisible to licensees, they become particularly important in arrangements involving collaboration by researchers at two or more universities. In such settings, institutions are often sensitive to perceived unequal treatment of collaborating researchers and must adjust their revenue-sharing policies to account for differing expectations.

Accolades and Criticisms

The Bayh-Dole Act and the university technology transfer structure it formed have generated numerous accolades and criticisms. Proponents of the Act contend that its encouragement of the patenting and licensing of federally funded research has provided an effective framework for federal technology transfer, yielding economic benefits not just for universities and private industry, but for the U.S. economy as a whole.[69] A 2002 article in the *Economist* famously referred to the Bayh-Dole Act as "possibly the most inspired piece of legislation to be enacted in America over the past half-century."[70] The Biotechnology Industry Organization reported that, in the period from 1996 to 2007, university licensing to industry created over 279,000 jobs and contributed to over $457 billion in industry output.[71] According to the former president of the Association of University Technology Managers, during the years 2000 to 2008 universities signed 41,598 license and option agreements with industry and filed 83,988 patent applications.[72]

Despite these glowing numbers, critics of the Act argue that the technology transfer system is inefficient and detrimental to the mission and norms of university research.[73] Relatively few of the patent applications filed by universities resulted in licensing agreements with industry, and fewer still resulted in large revenues, with only 0.5 percent of licensing agreements over the past 20 years exceeding $1 million in royalty income.[74] In 2005, only 25 universities reported more than $10 million in licensing revenue—a small amount in comparison to the research expenditures at many universities.[75] For most universities, revenue from licensing barely covers the cost of staff and legal expenses associated with the process.[76] Furthermore, some critics contend that the race to patent university research, and the revenue generated by university-owned patents, have caused many universities to shift their focus from basic research to commercial development.[77] This shift, they argue, has led to a reduction in nonremunerative basic research, a stifling of the free flow of ideas that previously characterized scientific inquiry, and an inappropriate linkage, if not an outright conflict of interest, that afflicts not only academic institutions but also individual investigators who stand to gain substantial financial rewards from the commercial exploitation of their laboratory research.[78]

To date, there is little definitive empirical evidence supporting either position.[79] Indisputable, however, is the fact that universities continue to develop innovations across a broad range of technologies, to obtain patent protection for those innovations (over 3,000 U.S. patents per year),[80] and to license those patents to the private sector for commercial application.

OTHER UNIVERSITY POLICY CONSIDERATIONS

Despite the frequent appearance of universities in the modern research-and-development landscape, universities are fundamentally different than corporate technology developers. Universities operate on a not-for-profit basis, their missions are directed primarily toward research and education, and they are populated largely by academics, scientists, and students. These unique characteristics distinguish university-based bioinformatics research, technology development, and exploitation, and result in policies and practices that are significantly different than those found in commercial settings.

The Research Exemption

A Narrow(ed) Exemption

A university's ability to carry on research freely and without impediment is fundamental to its mission. A decade ago, it was widely believed that academic research in the United States could be conducted without threat of patent infringement, on the basis that pure research does not infringe the exclusive rights of a patent holder (i.e., the rights to make, use, and sell a patented article and to perform a patented process).[81] This assumption was severely undermined by the Federal Circuit's 2002 decision in *Madey v. Duke University*.[82] In that case Professor Madey, a senior academic researcher, sued Duke, his former employer, for infringing several patents that Madey held in his own name. The alleged infringement involved Duke's continuing use of experimental laser equipment developed by Madey during and before his tenure at Duke. Duke asserted, among other things, that its use of the equipment had no commercial application and was directed solely to its nonprofit research mission. The court, while recognizing a limited judicial "experimental use" exemption from patent infringement, held that this exemption should be interpreted narrowly to exclude from infringement only activities that are carried out "for amusement, to satisfy idle curiosity, or for strictly philosophical inquiry."[83] Duke, it held, did not meet this standard, as its research was intended to further institutional business objectives such as educating students, improving its academic standing, and attracting research grants, students, and faculty.[84] As numerous commentators have observed, the *Madey* court's narrow reading of the experimental use exemption effectively eliminates its use in all but the most extreme cases, and does little to protect the research activities of any modern research university.[85]

The Limited Reach of Madey

It is worth noting two significant categories of institutions to which the Federal Circuit's narrow experimental use exemption does not apply. First, due to the territorial nature of patent law, the *Madey* decision applies only in the United States. Other jurisdictions, including the United Kingdom, have recognized infringement exemptions for experimental use that are still believed to protect most noncommercial academic research.[86] In the aftermath of *Madey*, some commentators have called for the U.S. Congress to enact a broad patent immunity for research and experimental activity.[87] To date, Congress has acted only incrementally by exempting from infringement experimentation conducted in furtherance of regulatory submissions for drugs and veterinary products.[88] A more general legislative experimental use exemption does not currently appear to be on the horizon.[89]

In addition to non-U.S. institutions, state-sponsored colleges and universities within the United States, which are immune from suit in federal court under the 11th Amendment of the U.S. Constitution, cannot be sued for patent infringement. Accordingly, state-operated research institutions such as the University of Michigan, the University of Wisconsin–Madison, the University of Florida, and the entire University of California system, each of which applies for and is awarded large numbers of patents every year, are themselves immune from patent infringement claims under current Supreme Court interpretation of the 11th Amendment.[90] While there have been calls to eliminate this apparent windfall to state universities,[91] such legislative proposals have not yet been successful. It is thus private U.S. universities that bear the brunt of the limited experimental use exemption.

Preserving Research Use through Contract

Given the limited scope of the experimental use exemption, private universities in the United States must conduct their research activities in the shadow of potential patent infringement. While there is evidence that many academic scientists ignore or are unaware of potential patent risks,[92] other evidence also suggests that potential patent claims may deter research in certain areas.[93] If nothing else, university TLOs and legal offices have become significantly more aware of potential infringement issues. According to one report, the University of Iowa, in attempting to clear the research being conducted at a single laboratory studying rare ocular disorders, unearthed 71 different entities of concern and spent $24,000 on background checks and queries to patent holders.[94]

Absent a change in the judicial interpretation of the experimental use exemption, universities are likely exposed to some level of risk from infringement of third-party patents. Such exposure may be unavoidable for the university that wishes to conduct research at the cutting edge of science. What is avoidable, however, is the risk that universities face from the patents on their own inventions. There have been recent examples of universities that, whether through inadvertence or carelessness, licensed inventions for exclusive use by industrial partners, thereby blocking any further use or development by the university laboratory that originated them.[95]

To avoid such situations, most universities now require standard language in all license agreements that reserve the university's right to exploit licensed inventions for their own noncommercial research and educational purposes.[96] In 2007, a group of major research universities together with the Association of American Medical Colleges (AAMC) released a document setting forth nine principles relevant to the licensing of academic technology "in the public interest and for society's benefit" (the Nine Points Document).[97] The first of these principles calls for universities not only to retain through their licensing agreements the right to practice licensed inventions, but also to extend such rights to any other nonprofit or governmental organization.[98] The Nine Points Document goes so far as to suggest that even research sponsored by commercial entities should be permitted, so long as it is conducted by a nonprofit entity. Ordinarily, such a reservation of rights would benefit a third-party university only if the licensing university granted it a license under the relevant patents. However, the Nine Points Document, which has now been endorsed by over 90 universities, also suggests an approach whereby any industrial licensee would contractually agree not to enforce a licensed patent against any university or other nonprofit institution.[99] Additionally, some funding organizations such as the NIH and the California Institute for Regenerative Medicine, are encouraging the creation of contractually based research exemptions for noncommercial research.[100] Should such contractual language be adopted widely by universities, a broad, contractually constructed experimental use exemption could emerge where Congress has failed to recognize one.

Publication and Data Release

While university administrators and technology transfer officers may be increasingly concerned with maximizing licensing and royalty revenue for their institutions, the currency of academic researchers is, and always has been, publication. The quantity and quality of a scientist's publications have been among the most important factors used in assessing the quality of his or her research, advancing his or her career, and determining his or her stature within the scientific community.[101] It is not surprising, then, that most university licensing and sponsored research agreements expressly reserve the right of university researchers to publish the results of their work in scholarly or scientific journals. If the work is being performed on behalf of a corporate sponsor or is likely to contain trade secrets of an industrial collaborator, it is not unusual for the agreement to require the university to provide a draft of any publication to the sponsor or collaborator in advance of publication, and to allow a period (usually 30 to 60 days) during which the sponsor or collaborator may suggest changes to preserve the ability to file patent applications and/or to redact trade secrets and confidential information.

A scientific publication typically includes a brief presentation of significant experimental findings, often made in summary or tabular fashion, together with the scientist's analysis and conclusions based upon those findings.[102] While the published data are usually essential to support the scientist's analysis, the data reported in a journal article seldom represent the entirety of the "raw" data collected or observed by the scientist, and are typically only a small fraction of the full data set. Over the past decade, however, an increasing number of scientific journals have required that authors make the data supporting their published claims available to readers upon request.[103] In certain fields such as genomics, government funding agencies routinely require the deposit of raw data sets into public databases.[104] It is likely that the trend toward broad sharing of, and access to, scientific data will continue through a combination of journal requirements, funding obligations, and academic agreements.

Socially Responsible Licensing

For the past decade there has been mounting public pressure to expand the availability of patented technologies, particularly so-called essential medicines, in the developing world. When the HIV anti-retroviral drug Zerit®, developed and patented by researchers at Yale University, became a critical part of the standard AIDS treatment regimen, Yale students and faculty, together with the popular press, exerted sufficient pressure on the university's exclusive licensee Bristol-Myers Squibb to persuade the company in 2001 to make the drug available at nominal cost to AIDS sufferers in Africa.[105] Since the Zerit® episode, an increasing number of universities have declared their support for such humanitarian or "socially responsible" licensing.[106] The 2007 Nine Points Document refers explicitly to the university's "social compact with society" and urges universities to structure their licensing arrangements so as to ensure that underprivileged populations have access to medical innovations.[107] In 2009 a group of six major research universities endorsed an even stronger statement committing that their intellectual property would not "become a barrier to essential health-related technologies needed by patients in developing countries."[108]

Potential licensing structures that might emerge through the trend toward socially responsible licensing by universities, as suggested by the experience of essential medicines, include (a) excluding developing countries from exclusive license grants, (b) requiring licensees to grant sublicenses to local producers in developing countries, (c) retaining university private march-in rights if products are not made suitably accessible in developing countries, and (d) prohibiting the filing of corresponding patent applications in developing countries.[109] Other contractual approaches that may achieve socially responsible goals include university patent pledges and nonassertion covenants such as a patent commons (described above under the heading "Patent Pools and Commons" in the section titled "Modes of University Research and Technology Transfer"), as well as the contribution of patents to socially oriented patent pools along the lines of the newly formed UNITAID[110] pool for essential medicines.[111]

CONCLUSION

Research universities have traditionally been catalysts for technological innovation and are likely to generate significant advances in bioinformatics technology for decades to come. However, unlike commercial enterprises, universities are subject to significant limitations and obligations arising from federal funding requirements, statutory regimes such as the Bayh-Dole Act, and the dictates of their nonprofit charters. It is important to keep these particular characteristics of universities and university research in mind when considering any collaboration, license, or sponsorship arrangement with them. If appropriate consideration is given to these characteristics, however, substantial benefits may be derived for industry, academia, and society as a whole.

NOTES

1. Committee on Management of University Intellectual Property, Managing University Intellectual Property in the Public Interest 16 (Stephen A. Merrill & Anne-Marie Mazza eds. 2010) [hereinafter University Intellectual Property].

2. *See generally* Ass'n of University Technology Managers, *AUTM U.S. Licensing Activity Survey FY 2008* (2010) [hereinafter AUTM FY2008]; Y. Friedman, *Biotech's U.S. Birth*, Sci. Am.—Worldview 54–57 (2009); T. D'Amato, J. L. Gilroy & S. Oldach, *From the Classroom to the Boardroom—How Universities Can Become the Flywheel for Economic Growth*, Intell. Prop. Today 22–25 (2009).

3. Ass'n. of University Technology Managers, *AUTM U.S. Licensing Activity Survey FY 2009* 13 (2011) [hereinafter AUTM FY2009]; AUTM Press Release, New Data Reveal University Startup Creation, Licensing Activity Strong Despite Economic Downturn (Oct. 11, 2012), http://www.innovations-report.com/html/reports/economy_finances/data_reveal_university_startup_creation_licensing_162976.html.

4. D. Zuhn, *IPO Releases List of Top 300 Patent Holders for 2011*, http://www.patentdocs.org/2012/06/ipo-releases-list-of-top-300-patent-holders-for-2011.html (last visited Aug. 8, 2012).

5. National Science Board, *Science and Engineering Indicators 2012* (Aug. 12, 2012), http://www.nsf.gov/statistics/seind12/pdf/seind12.pdf.

6. *See, e.g.*, Duke University Dept. of Biostatistics and Bioinformatics, http://biostat.duke.edu/research (last visited Aug. 12, 2012); Cornell University Department of Biological Statistics and Computational Biology, http://www.bscb.cornell.edu/ (last visited Aug. 12, 2012); University of California, Berkeley Center for Computational Biology, http://qb3.berkeley.edu/ccb/ (last visited Aug. 12, 2012).

7. AUTM FY2008, *supra* note 2, at 18.

8. *See* Peter Lyster, *Update on Biomedical Computation at NIH*, Symbios (Aug. 9, 2012), http://biomedicalcomputationreview.org/sites/default/files/spr2010-guest-editorial.pdf.

9. AUTM FY2009, *supra* note 3, at 13.

10. *See* E. Boyd & L. Bero, *Assessing Faculty Financial Relationships with Industry: A Case Study*, 284 J. Am. Med. Ass'n 2209–10 (2000).

11. Vanderbilt Univ. v. ICOS Corp., 601 F.3d 1297 (Fed. Cir. 2010).
12. AUTM FY2008, *supra* note 2, at 15–16.
13. AUTM FY2009, *supra* note 3, at 32–33.
14. *Id.* at 25.
15. *See* University Intellectual Property, *supra* note 1, at 24.
16. *See* M. Gordon, *University Controlled or Owned Technology: The State of Commercialization and Recommendations*, Les Nouvelles—J. Licensing Execs. Soc'y Int'l 152–63 (2004).
17. Wisconsin Alumni Research Foundation, *Our History*, http://www.warf.org/about/index.jsp?cid=26 (last visited Aug. 18, 2010).
18. *Id.*
19. *Id.*
20. *See* AUTM FY2009, *supra* note 3, at 13; University Intellectual Property, *supra* note 1, at 54.
21. *See* AUTM FY2008, *supra* note 2, at 39.
22. *See* J. Contreras & K. Eavis, *The Dizzying Rise of University Spinouts*, Tornado Insider 25–27 (2002).
23. *See* Daniel S. Greenberg, Science for Sale: The Perils, Rewards, and Delusions of Campus Capitalism ch. 10 (2007); University Intellectual Property, *supra* note 1, at 36–38.
24. *See generally* E. Elhauge, *Do Patent Holdup and Royalty Stacking Lead to Systematically Excessive Royalties?*, 4 J. Competition L. & Econ. 535 (2008); M. Lemley & C. Shapiro, *Patent Holdup and Royalty Stacking*, 85 Tex. L. Rev. 1991 (2007); C. Shapiro, *Navigating the Patent Thicket: Cross-Licenses, Patent Pools and Standard Setting, in* 4 Innovation Policy and the Economy 119 (Adam B. Jaffe, Josh Lerner & Scott Stern eds., 2004).
25. *See* Lemley & Shapiro, *supra* note 24; Shapiro, *supra* note 24.
26. *See generally* Institute of Medicine & Nat'l Research Council, *Large-Scale Biomedical Science* 234–37 (2003).
27. *See* National Science Foundation, *Survey of Research and Development Expenditures at Universities and Colleges (Table 1)*, http://www.nsf.gov/statistics/nsf10311/pdf/tab1.pdf (last visited Aug. 18, 2010) [hereinafter National Science Foundation Survey]; *see also* D.C. Mowery, *The Bayh-Dole Act and High-Technology Entrepreneurship in US Universities: Chicken, Egg, or Something Else?, in* 16 Advances in the Study of Entrepreneurship, Innovation and Economic Growth vol. 46 (G.D. Libecap ed., 2005).
28. AUTM FY2008, *supra* note 2, at 19; National Science Foundation Survey, *supra* note 27.
29. *See* R.S. Eisenberg, *Public Research and Private Development: Patents and Technology Transfer in Government-Sponsored Research*, 82 Va. L. Rev. 1663 (1996); University Intellectual Property, *supra* note 1, at 19.
30. *See* Greenberg, *supra* note 23, at 52; University Intellectual Property, *supra* note 1, at 24.
31. 35 U.S.C. §§ 200–212.
32. *See* University Intellectual Property, *supra* note 1, at 19–20.
33. 35 U.S.C. §§ 202–203.
34. 35 U.S.C. § 202(c)(1); 37 C.F.R. § 401.14(c)(1).
35. 37 C.F.R. § 401.14(f)(2).
36. *See, e.g.*, Bd. of Trs. of the Leland Stanford Jr. Univ. v. Roche Molecular Sys., Inc., 583 F.3d 832, 841–42 (Fed. Cir. 2009) (interpreting clauses of such an agreement from Stanford).
37. iEdison.gov, https://s-edison.info.nih.gov/iEdison/ (last visited Nov. 5, 2012).
38. 35 U.S.C. § 202(c)(5); 37 C.F.R. § 401.8(b).
39. 35 U.S.C. § 202(c)(1); 37 C.F.R. § 401.14(d)(1).
40. TM Patents v. IBM, 121 F. Supp. 2d 349 (S.D.N.Y. 2000); Thermalon Indus. Ltd. v. United States., 34 Fed. Cl. 414 (1995). For a more detailed analysis of these cases, *see* S.D. Locke, *Patent Litigation Over Federally Funded Inventions and the Consequences of Failing to Comply with Bayh-Dole*, 8 Va. J.L. & Tech. 3 (2003).
41. 37 C.F.R. § 401.14(e)(1).
42. 35 U.S.C. § 202(c)(2); 37 C.F.R. § 401.14(c)(2).
43. 35 U.S.C. § 202(c)(3); 37 C.F.R. § 401.14(c)(3).
44. 37 C.F.R. § 401.14(d)(2)–(3).
45. *See* AUTM FY2009, *supra* note 3, at 25.
46. 37 C.F.R. § 401.14(f)(3).
47. Bd. of Trs. of the Leland Stanford Jr. Univ. v. Roche Molecular Sys., Inc., 131 S. Ct. 2188 (2011).
48. *Id.* at 2192.
49. *Id.*
50. *Id.* at 2195.
51. *Id.* at 2192:

> Since 1790, the patent law has operated on the premise that rights in an invention belong to the inventor. The question here is whether the University and Small Business Patent Procedures Act of 1980—commonly referred to as the Bayh–Dole Act—displaces that norm and automatically vests title to federally funded inventions in federal contractors. We hold that it does not.

52. *Id.* at 2196–97.
53. *Id.* at 2199.
54. 35 U.S.C. § 202(c)(4); 37 C.F.R. § 401.14(c)(3).
55. 35 U.S.C. § 203; 37 C.F.R. §§ 401.6, 401.14(j).
56. *See* J.H. Raubitschek, & N.J. Latker, *Reasonable Pricing—A New Twist for March-In Rights under the Bayh-Dole Act*, 22 SANTA CLARA COMPUTER & HIGH TECH. L.J. 157 (2005).
57. National Institutes of Health, *Determination in the Case of Petition of CellPro, Inc.*, http://www.nih.gov/news/pr/aug97/nihb-01.htm (last visited Aug. 18, 2010).
58. *See* Raubitschek & Latker, *supra* note 56, at 158.
59. National Institutes of Health, *In the Case of NORVIR(R) Manufactured by Abbott Laboratories* (Aug. 18, 2010), http://OTT.od.nih.gov/Reports/March-In-Norvir.pdf.
60. *See* Richard Knox, *With a Life-Saving Medicine in Short Supply, Patients Want Patent Broken*, http://www.npr.org/blogs/health/2010/08/04/128973687/with-a-life-saving-medicine-in-short-supply-patients-want-patent-broken (last visited Sept. 29, 2010).
61. *Id.*
62. A copy of the petition is available at http://patentlawyersite.com/files/Download/Fabrazyme_Petition_5_0.doc (last visited Sept. 29, 2010).
63. Health of the Director, *Determination in the Case of Fabrazyme Manufactured by Genzyme Corporation*, Dec. 1, 2010, at 9.
64. 35 U.S.C. § 204; 37 C.F.R. § 401.14(i).
65. *See* S. Boettiger & A.B. Bennett, *Bayh-Dole: If We Knew Then What We Know Now*, 24 NATURE BIOTECHNOLOGY 320, 320 (2006).
66. 35 U.S.C. § 202(c)(7)(A); 37 C.F.R. § 401.14(k)(1).
67. 35 U.S.C. § 202(c)(7)(B)–(C); 37 C.F.R. § 401.14(k)(2)–(3).
68. Stanford University Office of Technology Licensing, *OTL's Standard Operating Procedure*, http://otl.stanford.edu/inventors/inventors_process.html (last visited Aug. 18, 2010); N. Bhakuni, *From Conception to Commercialization—University Technology Transfer Practices in the United States*, LES NOUVELLES—J. OF THE LICENSING EXECS. SOC'Y INT'L 62 (2006).
69. *See, e.g.*, B. Bayh, J.P. Allen & H.W. Bremer, *Universities, Inventors, and the Bayh-Dole Act*, 3 LIFE SCI. L. & INDUS. 1 (2009); H.R. Rep. No. 106-129, at 6 (2000); *Innovation's Golden Goose*, ECONOMIST, Dec. 14, 2002; E. McConathy, *A Report to Congress on the Success of the Bayh-Dole Act in 25-Year Hindsight*, LICENSING J. 21 (2007).
70. *Innovation's Golden Goose*, ECONOMIST, Dec. 14, 2002, at 3.
71. Biotechnology Industry Organization, *Final Report to the Biotechnology Industry Organization* (Sept. 17, 2010), http://www.bio.org/ip/techtransfer/BIO_final_report_9_3_09_rev_2.pdf.
72. Arundeep S. Pradhan, *Defending the University Tech Transfer System*, BUSINESSWEEK (Feb. 19, 2010), http://www.businessweek.com/smallbiz/content/feb2010/sb20100219_307735.htm.
73. *See* UNIVERSITY INTELLECTUAL PROPERTY, *supra* note 1, at 22; GREENBERG, *supra* note 23; J. WASHBURN, UNIVERSITY, INC: THE CORPORATE CORRUPTION OF HIGHER EDUCATION (2005).
74. *See* UNIVERSITY INTELLECTUAL PROPERTY, *supra* note 1, at 27; and GREENBERG, *supra* note 23, at 60.
75. *See* GREENBERG, *supra* note 23, at 62–63.
76. *See* UNIVERSITY INTELLECTUAL PROPERTY, *supra* note 1, at 27; and GREENBERG, *supra* note 23, at 62.
77. *See* K.J. Strandburg, *Curiosity-Driven Research and University Technology Transfer, in* 16 ADVANCES IN THE STUDY OF ENTREPRENEURSHIP, INNOVATION AND ECONOMIC GROWTH 111–12 (G.D. Libecap ed., 2005).
78. *See* GREENBERG, *supra* note 23; A.K. Rai & R.S. Eisenberg, *Bayh-Dole Reform and the Progress of Biomedicine*, 66 LAW & CONTEMP. PROBS. 289 (2003); WASHBURN, *supra* note 73. *But see* UNIVERSITY INTELLECTUAL PROPERTY, *supra* note 1, at 36–41.
79. *See generally* Charles R. McManis & Sucheol Noh, *The Impact of the Bayh-Dole Act on Genetic Research and Development: Evaluating the Arguments and Empirical Evidence to Date* (Sept. 21, 2010), http://law.wustl.edu/CLIEG/documents/mcmaniscommercializinginnovationpaper.pdf.
80. *See* AUTM FY2008, *supra* note 2; AUTM FY2009, *supra* note 3, at 25.
81. *See, e.g.*, Boettiger & Bennett, *supra* note 65, at 321.
82. Madey v. Duke Univ., 307 F.3d 1351 (Fed. Cir. 2002).
83. *Id.* at 1362.
84. *Id.*
85. *See* J.L. Curry & B.E. O'Connor, *University Research—A New Defense Under the Patent Law*, 12 J. INTELL. PROP. L. 29 (2004); A.C. Yancey & C.N. Stewart, Jr., *Are University Researchers at Risk for Patent Infringement?*, 25 NATURE BIOTECHNOLOGY 11 (2007); Boettiger & Bennett, *supra* note 65, at 321; *see also* J. Carter-Johnson, *Unveiling the Distinction Between the University and Its Academic Researchers: Lessons for Patent Infringement and University Technology Transfer*, 12 VAND. J. ENT. & TECH. L. 473 (2010) (arguing that most research in a university is conducted by practically

independent research scientists who may in some instances still have acess to the narrow research exemption). *But see* E. A. Rowe, *The Experimental Use Exception to Patent Infringement: Do Universities Deserve Special Treatment?*, 57 Hastings L.J. 921 (2006) (arguing that the Federal Circuit did not narrow the experimental use defense, and that its scope is appropriate).

86. *See* Curry & O'Connor, *supra* note 85, at 33.

87. *See, e.g.*, *id.* at 36–37.

88. Hatch-Waxman Act, 35 U.S.C. § 271(e)(1) (2003). *See* Merck KGaA v. Integra Lifesciences Ltd., 545 U.S. 193 (2005) (clarifying the limited scope of the statutory exemption).

89. Although there currently is no general statutory research exemption, if the particular technology being developed is regulated by the Food and Drug Administration (FDA) a statutory research exemption specific to research related to FDA submissions may apply. 35 U.S.C. § 271(e) (2006).

90. Florida Prepaid Postsecondary Education Expense Board v. College Savings Bank, 527 U.S. 627 (1999).

91. *See* T.D. Quigley, *Commercialization of the State University: Why the Intellectual Property Protection Restoration Act of 2003 Is Necessary*, 152 U. Pa. L. Rev. 2001 (2004).

92. *See* J.P. Walsh, A. Arora & W.M. Cohen, *Effects of Research Tool Patents and Licensing on Biomedical Innovation in Patents, in* Patents in the Knowledge-Based Economy 285 (W.M. Cohen & S.A. Merrill eds., 2003) (finding, based on interviews and archival data, little evidence that academic research has been impeded by concerns about patents on research tools); *see also* J.P. Walsh, W.M. Cohen & C. Cho, *Where Excludability Matters: Material versus Intellectual Property in Academic Biomedical Research*, 36 Res. Pol'y 1184 (2007).

93. *See* J.F. Merz et al., *Diagnostic Testing Fails the Test: The Pitfalls of Patents are Illustrated by the Case of Hemochromatosis*, 415 Nature 577 (2002); M.K. Cho et al., *Effects of Patents and Licenses on the Provision of Clinical Genetic Testing Services*, 5 J. Molecular Diagnosis 3 (2003).

94. *See* B. Wysocki Jr., *A Laser Case Sears Universities Right to Ignore Patents*, Wall St. J., Oct. 11, 2004, at A1.

95. *See* Yancey & Stewart, *supra* note 85, at 1226–27.

96. *See* Boettinger & Bennett, *supra* note 65, at 321.

97. *In the Public Interest: Nine Points to Consider in Licensing University Technology*, http://www.autm.net/Nine_Points_to_Consider.htm (last visited Aug. 18, 2010) [hereinafter Nine Points Document].

98. *Id.*

99. *Endorse the Nine Points to Consider: Current Signatories*, https://www.autm.net/source/NinePoints/ninepoints_endorsement.cfm (last visited Aug. 10, 2012).

100. *See* P. Lee, *Contracting to Preserve Open Science: Consideration-Based Regulation in Patent Law*, 58 Emory L.J. 889, 920–38 (2009).

101. *See* R.K. Merton, Priorities in Scientific Discovery (1957), *reprinted in* 286 The Sociology of Science 316 (N.W. Storer ed. 1973); J. G. Mills III et al., Patent Law Fundamentals 2–50 (Thomas West ed., 2009).

102. *See* R.S. Eisenberg, *Patents and Data-Sharing in Public Science*, 15 Indus. & Corp. Change 1013, 1024 (2006).

103. *See, e.g.*, *Guide to Publication Policies of the Nature Journals*, http://www.nature.com/authors/gta.pdf (last visited Aug. 18, 2010) ("[A] condition of publication in a Nature journal is that authors are required to make materials, data and associated protocols promptly available to readers without preconditions.") (emphasis removed).

104. *See* Jorge L. Contreras, *Prepublication Data Release, Latency, and Genome Commons*, 329 Sci. 393, 393 (2010).

105. *See* A.J. Stevens & A.E. Effort, *Using Academic License Agreements to Promote Global Social Responsibility*, 85 Les Nouvelles—J. of Licensing Execs. Soc'y Int'l 86 (2008).

106. *See* University Intellectual Property, *supra* note 1, at 76.

107. *See* Nine Points Document, *supra* note 97, at 8.

108. *Statement of Principles and Strategies for the Equitable Dissemination of Medical Technologies, available at* http://www.autm.net/Content/NavigationMenu/TechTransfer/GlobalHealth/statementofprinclipies.pdf (last visited Aug. 18, 2010) (statement endorsed by Harvard University, Yale University, Brown University, Boston University, the University of Pennsylvania, Oregon Health & Science University, and AUTM).

109. *See, e.g.*, Stevens & Effort, *supra* note 105, at 91; Nine Points Document, *supra* note 97, at 8.

110. UNITAID is a global health initiative to provide funding for greater access to treatments and diagnostics for HIV/AIDS, malaria, and tuberculosis in low-income countries. http://www.unitaid.eu/who/about-unitaid.

111. UNITAID, http://www.unitaid.eu/en/The-Medicines-Patent-Pool-Initiative.html.

CHAPTER 10

Technical Standards and Bioinformatics

Jorge L. Contreras*

INTRODUCTION

Standards are ubiquitous in today's technology-driven marketplace.[1] Some standards mandate minimum requirements for product safety, some seek to reduce environmental impact, and others specify minimum levels of information that must be provided to consumers. "Interoperability standards," which dominate the information, computing, and telecommunications (ICT) sector, specify the manner in which products and services offered by different vendors interact with one another.[2] A number of these standards, including WiFi, USB, CD, DVD, PDF, and HTML, have become household terms, and thousands of others ensure that a vast array of products and services connect and communicate seamlessly in a manner that is largely invisible to the consumer.

Standardization in the ICT sector, however, has not been achieved without a substantial cost. Over the past two decades, the industry has been plagued by lawsuits brought by participants in the standards-development process, and by government regulators and affected third parties. Two types of claims generally arise in standards-related litigation: process-abuse claims and patent hold-up claims. Process-abuse claims involve claims alleging that the standards process has been abused to put one or more companies at a disadvantage, and also include claims that a participant in the standards-development process has improperly asserted its patents against an implementer of the standard. Standards-development organizations in the ICT sector have responded to these claims by promulgating rules and policies of increasing sophistication, both to specify procedures designed to avoid abusive activity and to accommodate the requirements of participants who control significant patent assets.

The explosion of bioinformatics research over the past two decades has led to a surge of interest in the development of interoperability and compatibility standards for

*This chapter is adapted from Jorge L. Contreras, *Implementing Procedural Safeguards for the Development of Bioinformatics Interoperability Standards*, 39 N. Ky. L. Rev. 87 (2012).

bioinformatics applications. These standards range from genome annotation and controlled vocabularies (ontologies) to data formats and search engine integration. A variety of organizations are involved in standards-development activities. Involved organizations range from large, established standards bodies such as the Institute for Electrical and Electronics Engineers (IEEE)[3] and the Worldwide Web Consortium (W3C),[4] to broad-based bioinformatics industry associations such as the Asia-Pacific Bioinformatics Network[5] and the European Bioinformatics Institute (IEBI),[6] to narrowly focused efforts such as the Proteomics Standards Initiative[7] and the Functional Genomics Investigation Ontology project.[8]

To date, the bioinformatics field has been blissfully free of the standards litigation that has plagued the ICT industry. But with the increasing adoption of standards by bioinformatics researchers and vendors, the issues faced by ICT standards groups will become increasingly relevant to bioinformaticists. I conducted an industry-wide study of bioinformatics standards-development activities and reviewed the policies and procedures adopted by each standards-development organization. The results of this study show that the majority of bioinformatics standards-development efforts are relatively informal and unstructured. In many cases, these organizations either lack written policies entirely or adopt vague, aspirational statements regarding a desire that materials produced be "open" and publicly available. Accordingly, this chapter recommends that bioinformatics standards-development organizations carefully review their existing policies and procedures and revise them to the extent that they fail to address key points regarding process and intellectual property.

STANDARDS AND STANDARD SETTING: A BRIEF OVERVIEW

Types of Standards

Standards serve a variety of purposes and functions. Some standards specify requirements intended to protect public health and safety, to preserve the environment, and to prevent fraud and other abuses. These include standards relating to food and drugs, air and water quality, hazardous materials, construction, transportation, handling of personal data, and the like. Other standards signify a specified level of quality, such as the 4C (cut, color, clarity, and carat) ratings for diamonds. Most relevant to bioinformatics, however, are interoperability or compatibility standards, which specify design features that enable data, materials, and services originated by different parties to operate together. Electrical outlets, for example, share a common design in the United States that enables any appliance to be plugged into any outlet anywhere in the country (though, as any frequent traveler knows, these standards vary dramatically from country to country). More complex, but equally illustrative, are the numerous networking (USB, WiFi, Bluetooth), Internet (TCP/IP, HTML, http), telecommunications (CDMA, GSM, LTE), and digital media (CD, DVD, BluRay) standards that enable devices manufactured by different vendors to interact with one another in a manner that is largely invisible to the consumer.

Mandatory and Voluntary Standards

At the highest level of generality, standards may be categorized as either (1) "mandatory," meaning that compliance is required by an external body, such as a governmental agency or a professional accreditation organization, or (2) "voluntary," meaning that compliance is not required, though it may be prudent or even necessary from a commercial standpoint. Mandatory standards are often adopted by governments to protect the health and welfare of their citizens. Most interoperability standards, however, are voluntary.

Standards Development Organizations[9]

Standards may be developed in a number of different environments. Individual companies may simply release a technology or format to the industry in the hope that it will become broadly adopted. Such "de facto" standards include Adobe Systems' portable document format (PDF) and Microsoft's ".doc" document format. Other standards may be developed by a group of companies and institutions working collaboratively within a standards-development organization (SDO). SDOs vary greatly in size and composition. Some SDOs might consist of just a few participants that collaborate on a focused set of technical specifications, sometimes for a single product. These are sometimes referred to as consortia or "special interest groups" (SIGs). Standards for many consumer electronics devices and digital media such as the DVD disc and player were developed by SIGs. Other SDOs are very large and encompass many different standardization activities at any given time. ASTM International, for example, is one of the largest SDOs and regularly develops standards in areas as diverse as electrical wiring, playground equipment, composite materials, unmanned aircraft, and nanotechnology.[10]

The work of individual SDOs is sometimes coordinated at the national and international levels. In the United States, SDOs can be accredited by the American National Standards Institute (ANSI) if they meet certain criteria of openness, due process, and transparency (essential requirements).[11] The standards developed by ANSI-approved SDOs can be designated as American National Standards, a status that both validates and promotes adoption.[12] More than 200 different SDOs are currently accredited by ANSI.[13] The International Organization for Standardization (ISO)[14] is a Geneva-based nongovernmental organization whose members include 163 national standards institutes from across the globe. ISO coordinates standards-development activities among its member and develops its own consensus standards through numerous committees.

Many federal agencies also adopt standards, either as part of their regulatory function or for their own use. The Federal Office of Management and Budget's guidelines for federal agencies require that agencies adopt suitable "voluntary consensus standards" in their procurement and regulatory activities to the extent not "inconsistent with law or otherwise impractical."[15] At an international level, the World Trade Organization's Uruguay Round Agreement on Technical Barriers to Trade requires that if recognized international standards exist with respect to a technical area, national governments developing standards in that area must adopt such international standards as the basis for their own national standards.[16]

THE BIOINFORMATICS STANDARDS LANDSCAPE

Bioinformatics research utilizes a broad range of technologies, including experimental apparatus such as microarrays, data analysis tools, and databases that store and allow sharing of experimental and analytic data.[17] Standards are required within each of these broad technology categories to enable data sharing, analysis, and the interoperability of different experimental platforms. To date, hundreds of standards relevant to bioinformatics applications have been developed in three broad categories: terminological artifacts, reporting requirements, and exchange formats.[18] Below is a brief description of these categories and a summary of some of the more prominent standardization efforts being undertaken in each.

Terminological Artifacts

A consistent and unambiguous vocabulary is required in order for different groups and applications to communicate about a wide array of organisms, experimental conditions, and study designs.[19] A "controlled vocabulary" offers a single set of terms with explicitly defined meanings within a particular field. An "ontology" creates relationships among these terms, often in hierarchical form, such as the familiar taxonomic classification for biological entities (kingdom, phylum, class, etc.).

One of the earliest and most mature bioinformatics ontologies is the Gene Ontology (GO), which has produced "a structured, precisely defined, common, controlled vocabulary for describing the role of genes and gene products in any organism."[20] The Gene Ontology Consortium, which developed the GO, began its work in 1998 as a joint project of research groups studying three different organisms (fruit fly, budding yeast, and mouse). As of 2009, the GO included more than 44.5 million annotated gene products.[21]

Numerous other ontology projects have arisen following the success of the GO.[22] These include the Systems Biology Ontology (SBO) originated by the European Bioinformatics Institute (EBI),[23] the Open Biological and Biomedical Ontologies (OBO Foundry),[24] and the Functional Genomics Investigation Ontology (FuGO) project.[25]

Reporting Requirements

The advent of microarray technology in the 1990s quickly led to the realization by the research community that standardized methods of reporting experimental data generated by microarray studies would be required. The Microarray Gene Expression Data Society (now the Functional Genomics Data Society) was formed in 1999 by EBI to develop "Minimum Information About a Microarray Experiment" (MIAME), a checklist specifying the information about every microarray experiment that should be reported in order to enable its proper validation, reproduction, and interpretation.[26]

Many other minimum information standards development efforts have followed the early success of MIAME.[27] Among others, these include specifications for minimum information in hybridization and immunohistochemistry experiments (MISFISHIE),[28] proteomics experiments (MIAPE),[29] molecular interactions (MIMIx),[30] and (meta)genome sequences (MIGS/MIMS).[31] The Minimum Information for Biological and Biomedical Investigations (MIBBI) Project collects and publishes information about various minimum information standards.[32]

Exchange Formats

A recent compilation of molecular biology databases lists 1,330 different data sources across the world.[33] In order for researchers to make use of data beyond their own laboratories, they require the ability to access and utilize data from disparate sources. The exchange of data among different software applications and databases has become commonplace in today's data-driven economy. Much of this exchange is accomplished using markup languages, or schema that enable the annotation of digital text in a manner that can be interpreted by a computer.[34] The most common markup language, HyperText Markup Language (HTML), is the dominant language for encoding web pages.[35] Markup languages work through the use of "tags" that delimit characteristics of the text they designate. For example, in HTML, the tags <B> and </B> cause the text between the tags to be displayed in boldface type. Extensible Markup Language (XML), developed by the W3C, is a flexible data format that enables users to create customized tags based on the specific types of data in which they are interested.[36] Hundreds of XML-based languages exist today, many of which are optimized for bioinformatics applications.[37] Among the best-known XML-based bio-focused languages are the Systems Biology Markup Language (SBML),[38] CellML for computational cell biology,[39] MicroArray Gene Expression Markup Language (MAGE-ML) for the exchange of microarray data,[40] and Biological Pathway Exchange (BioPAX) for the exchange of biological pathway data.[41]

Despite the widespread adoption of XML as the preferred standard for data exchange in the biosciences, some commentators have criticized XML as too limited and imprecise for the robust exchange of scientific data.[42] An alternative data exchange standard is the reference data format (RDF), also developed by W3C, which takes advantage of the so-called semantic web and offers developers greater freedom to define data relationships.[43] The Bio2RDF Project led by W3C has recently developed software and a website for accessing data from different public bioinformatics databases in RDF format.[44]

Closely related to markup languages are object models. Object models describe the relationships among computer programming "objects" of interest within a given discipline and can thus act as blueprints for the development of specialized markup languages.[45] Of particular interest in bioinformatics are the Microarray and Gene Expression Object Model (MAGE-OM),[46] the SysBio-OM developed by the National Institute of Environmental Health Sciences (NIEHS),[47] and the Functional Genomics Experiment Object Model (FuGE-OM).[48]

The object models and markup languages discussed above are essential for the exchange of data among disparate data sources and databases. Yet data exchange tools alone are not enough, and commentators find that the proliferation of incompatible data sources has led to increasing duplication of effort, poor interoperability, and loss of data.[49] To address these problems, an international consortium of database users and developers has begun work on a uniform set of defining attributes for biological databases called BioDBCore.[50]

COMPETITION AND ANTITRUST ISSUES FOR STANDARDS DEVELOPERS

One of the principal purposes of antitrust law is to prevent competitors from colluding in ways that are harmful to competition and the market. Such collusive behavior in the United States is typically prosecuted under Section 1 of the Sherman Act, which broadly prohibits agreements in restraint of trade.[51] In recent years, the Federal Trade Commission (FTC) has also exercised its authority under Section 5 of the FTC Act to prosecute unfair methods of competition and unfair trade practices that arise in the standards-setting context.[52] Because consensus-based technical standards are generally developed through collaborative processes that involve multiple competitors, having a basic understanding of antitrust law principles is critical to effective operation of any standards-development organization. This section summarizes some of the major antitrust and competition issues that can arise in the standards-setting context.

Improper Exclusion

SDOs must exercise caution if they wish to limit their membership or to expel a member after it has joined.[53] The exclusion of parties from the standard-setting process can be viewed as an attempt by the SDO members to gain an unfair advantage over the excluded parties. Accordingly, many SDOs do not restrict their membership and claim to be open to all interested participants.[54] Being "open," however, does not mean that an SDO must admit all participants on equal terms. It is not uncommon, for example, for SDOs to vary their membership fees based on a participant's size, revenue, or industry sector (e.g., producer, customer, government, academic, etc.).[55] Courts have also held that trade associations may establish reasonable criteria for membership, even if such criteria necessarily exclude some potential participants;[56] many important industry standards, such as the CD and DVD digital storage formats were developed by small, closed groups of industry participants. Although some limitations may be permitted, SDOs can avoid many allegations of improper exclusion by opening their membership as broadly as possible. Similar issues regarding exclusion arise when a group wishes to expel a member. In short, expulsion of a member is generally permissible when the member has violated the organization's rules or otherwise acted in an unlawful or unethical manner.[57]

Claims of anticompetitive exclusion can arise in connection with technologies as well as participants. Thus, even if a company is not excluded from participation in an SDO, it is possible for other SDO participants to collude to exclude that company's technology from a standard under consideration. Such claims were raised, for example, by Addamax Corporation in *Addamax Corp. v. Open Software Foundation Inc.*[58] Addamax was a developer of security software for the UNIX operating system. Open Software Foundation (OSF), a consortium of computer and software vendors, developed a version of UNIX that included a security package developed by one of Addamax's competitors. Addamax alleged, among other things, that OSF's exclusion of its security software from OSF UNIX amounted to unfair competition and a boycott of Addamax's products by the members of OSF. After lengthy litigation, the First Circuit rejected these claims, holding that Addamax failed to show that the eventual decline in sales of its security products was attributable to concerted action by OSF and its members.[59]

Improper Exchange of Information

If competitors exchange sensitive commercial information during standard-setting meetings, or if they otherwise use the standard-setting process as a vehicle for engaging in illegal collusion, the SDO can be challenged for facilitating this unlawful behavior. The clearest example of behavior exposing SDOs to potential antitrust risk is the sharing of information relating to product pricing or salaries.[60] But less blatant instances of information sharing may also lead to allegations of anticompetitive behavior. For example, a standard could facilitate collusion by making it easier for competitors to monitor and enforce illegal price-fixing agreements.[61] Likewise, a real estate SDO's bylaws and policies could facilitate collusion among members by prohibiting a broker from conveying a listing to his customers via the Internet without permission of the listing broker.[62]

Compliance Certification

In some cases, the members of an SDO may improperly influence the SDO's decisions regarding whether a competitor's products comply with a standard. The Supreme Court addressed this type of behavior in *American Society of Mechanical Engineers v. Hydrolevel Corp.*[63] In that case, a subcommittee chair of the American Society of Mechanical Engineers (ASME), an SDO with over 90,000 members, misrepresented that Hydrolevel's boiler control valves did not conform to a key provision of ASME's Boiler and Pressure Vessel Code, and were therefore unsafe. He allegedly made these misrepresentations because his employer competed with Hydrolevel in the boiler control valves market. When ASME staff further propagated this inaccurate portrayal, Hydrolevel alleged that it suffered significant harm in the market. The Supreme Court held that the SDO could be liable for these misrepresentations, as the weight of its reputation greatly enhanced the anticompetitive effect of its member's misrepresentation.[64]

But not all allegations that an SDO improperly failed to certify a product's compliance with its standards are successful. In *Consolidated Metal Products v. American Petroleum Institute*,[65] the plaintiff alleged that the American Petroleum Institute committed an antitrust violation by intentionally delaying certification of its products. The Fifth Circuit rejected this claim, holding that the plaintiff did not have a claim "simply because others refuse to promote, approve, or buy its products."[66]

Abuse of Process

Perhaps the most frequent antitrust charge against SDOs is the claim that one or more participants abused the process by which standards are set within the SDO, either to secure approval of a standard that they backed or to exclude the technology of their competitors from a standard. The Supreme Court famously addressed this type of abuse in *Allied Tube & Conduit v. Indian Head*.[67] In that case, the plaintiff, a manufacturer of polyvinyl chloride (PVC) products, requested that the National Fire Protection Association (NFPA) approve an amendment to the National Electrical Code that would have allowed the use of PVC in electrical wiring conduits. The defendant, a manufacturer of steel conduit, recruited 230 people to join NFPA for the sole purpose of voting against the amendment, which was ultimately defeated by a vote of 394 to 390. The jury found that the steel conduit manufacturers defeated the PVC standard solely for anticompetitive purposes, subverting the standards-development process and violating the antitrust laws. The Supreme Court agreed, stating that a participant in private standards-setting activities should not bias the process in a manner designed to restrain competition.

Policies and Precautions

In order to avoid the antitrust claims discussed above, as well as the undesirable conduct by participants that led to such claims, SDOs have increasingly adopted rules and policies that expressly prohibit anticompetitive behavior within the SDO context.[68] Such policies typically seek to accomplish the following general objectives:

a. to ensure that the standards-development process is open, fair, and transparent;
b. to ensure that membership in the SDO is open to all interested parties on a non-discriminatory basis;
c. to prohibit the sharing of competitive information within the context of SDO activities; and
d. to provide that compliance with standards is voluntary, and that participation in the SDO is not conditioned on use of the SDO's output.

While policies implementing these objectives cannot themselves prevent collusion or improper behavior by SDO participants, they can serve to raise awareness of antitrust issues within the SDO. More importantly, such policies can demonstrate (whether to potential plaintiffs or finders of fact) that the SDO has actively sought to curtail activity that could be construed as anticompetitive.

INTELLECTUAL PROPERTY AND STANDARDS

Standards often implicate and are affected by intellectual property. Indeed, standards themselves may specify patentable inventions, and the written embodiments of standards are generally protectable by copyright. Furthermore, the more complicated the technology that a standard specifies, the more likely the standard is to implicate patents owned by members of the SDO or by third parties. SDOs and implementers of technical standards are therefore likely to encounter numerous intellectual property issues outlined in this section.

Copyright in Standards

Technical standards typically take the form of written descriptions of how products or services should be designed, built, or operated.[69] Similar to written documents, standards are typically protected by copyright, meaning that they cannot be reproduced, displayed, or modified without permission of the copyright owner (often the SDO).[70] Many SDOs earn significant revenue from the sale of standards (some of which extend to hundreds of pages) and therefore warn against illegal copying and distribution,[71] though a number of major SDOs allow their standards to be downloaded and copied without charge.[72] The tension between copyright protection and the social utility of standards becomes particularly clear when a proprietary standard is adopted and referenced by a governmental agency, thereby becoming "the law."[73] Use of a copyrighted standard might become mandatory by statute or regulation, yet access to that standard can be controlled by the SDO that owns the copyright.

Such a situation led to litigation when the federal Health Care Financing Administration (HCFA)[74] required the use of the Current Procedural Terminology (CPT)[75] standard for Medicare and Medicaid reimbursement claims.[76] The copyright in this standard was owned by the American Medical Association (AMA), which granted HCFA "a non-exclusive, royalty free, and irrevocable license to use, copy, publish, and distribute" the standard.[77] In *Practice Management Information Corp. v. American Medical Ass'n*, Practice Management Information (PMI), an independent publisher, desired to publish the CPT standard.[78] Arguing that the CPT standard was now incorporated in federal regulation, PMI sought a declaratory judgment that the AMA's copyright in the CPT standard was no longer valid.[79] The U.S. Court of Appeals for the Ninth Circuit, however, declined to invalidate the AMA's copyright, holding instead that the AMA controls the copyrighted text of the CPT standard, even after it was adopted into law.[80] This approach is consistent with the Federal Office of Management and Budget's guidance to federal agencies, which states that agencies adopting voluntary consensus standards "must observe and protect the rights of the copyright holder."[81]

However, the Court of Appeals for the Fifth Circuit has taken a different approach with respect to the issue of copyright in legally mandated standards. In *Veeck v. Southern Building Code Congress Int'l*,[82] Peter Veeck, the operator of a nonprofit website, posted the local building codes of two Texas municipalities.[83] The codes were taken verbatim from Southern Building Code Congress International's (SBCCI) published Standard Building Code.[84] When Veeck published the codes on his website, SBCCI sued him for infringement of its copyrights.[85] The *en banc* Fifth Circuit, relying on precedent that copyright cannot prevent the reproduction and distribution of "the law," held that while SBCCI retains the copyright in its model codes, the codes may be reproduced and distributed freely once enacted into law.[86] Important to the court's reasoning was the fact that SBCCI's model codes were developed specifically for the purpose of being adopted by municipalities and incorporated into their local building codes. The court distinguished this case from one in which a governmental agency simply references an existing standard in its regulations. In such a case, the court suggested that it would not be unreasonable to permit the authoring SDO to continue charging for access to its standard, notwithstanding its incorporation in such regulations.[87]

The debate regarding incorporation of published standards into legislative material has recently gained renewed attention following a February 2012 petition by several academics and public interest advocates seeking to amend the authorizing regulations for the Federal Register to require that technical standards referenced in federal regulations be made freely available via the Internet.[88] The petition has generated significant debate, and responses to a subsequent Federal Register request for comments have both supported[89] and opposed[90] the proposed amendment.

Patenting and Standards

Patents Covering Standardized Technology

As noted in the previous section, standards are written descriptions of particular attributes of specified products and processes. Assuming that statutory requirements of utility, novelty, and nonobviousness are met, patents may be obtained on bioinformatics products and processes that conform with particular standards. Such a patent would ordinarily not be obtained by the SDO in which the standard was developed, as SDOs seldom develop complete products and almost never seek to patent their work. Rather, if such a patent were obtained, it would be obtained by a participant in the SDO or an outsider, and sometimes both. Thus, two general patent-related issues arise in the context of technology standardization: "patent stacking" and "patent hold-up."

Patent Stacking and Patent Pools

Patent or royalty "stacking," also referred to as a patent "thicket" or anti-commons, occurs when multiple entities each hold patents claiming aspects of a single standardized technology.[91] In order for a producer to implement the standardized technology in a product, it must obtain licenses from multiple parties, each acting independently and each seeking to maximize its gains.[92] The risk of stacking is that the sum of individual royalty demands may be excessive in relation to the overall value of the product, making the standardized product uncompetitive in comparison to products that do not conform with the standard.[93]

One method of addressing patent stacking is through the creation of a patent pool. In a patent pool, multiple patent owners contribute or license patents that are essential to the implementation of the standard to a common agent (sometimes one of the patent holders and sometimes a newly formed entity). Licensees are charged a single royalty to practice the entire group of patents, and net revenues are allocated among the pool participants in accordance with a predetermined formula. Such pools have been used effectively in consumer electronics standards such as the MPEG audio compression format,[94] the DVD video compression format,[95] and third generation wireless communications standards.[96] In each of these instances, the U.S. Department of Justice approved the proposed patent pool, so long as it possessed certain features that would reduce potentially anticompetitive effects.[97] For example, pools must contain only patents "essential" to the implementation of the standard (as the inclusion of patents on substitute technologies could lessen competition among technical alternatives). Furthermore, licensees must have the freedom to obtain licenses to the patents independently from the pool, and licensing of the pooled patents must be conducted on a nondiscriminatory basis. To the extent that the patent pool owners require licensees to "grant back" licenses to them, such grant-back licenses must cover only patents that are themselves essential to implementation of the standard.

It is important to note that the usefulness of a patent pool may be limited if not all holders of essential patents participate. This situation arose recently in connection with the Federal Communication Commission's (FCC) mandatory ATSC digital television standard.[98] Though several holders of patents essential to the ATSC standard formed a patent pool, one patent holder, Japan's Funai Electric Company, did not join. Instead, Funai sought to charge royalties for a single patent at a rate allegedly equal to the rate charged by the entire ATSC pool. Vizio, Inc., a U.S. television manufacturer, refused to pay this royalty, causing Vizio to seek relief from the FCC. Though the matter was ultimately rendered moot when the Federal Circuit held that Vizio did not infringe the asserted patent,[99] the dispute highlights the limited benefits of patent-pooling arrangements that do not include all holders of standards-essential patents.

Patent Ambush

The second major patent issue that can arise in the standards context is patent ambush. Patent ambush occurs when a patent holder seeks to assert a previously unidentified patent against implementers of a standard after the standard has been adopted (either by an SDO or a governmental agency).[100] If patent ambush occurs after the industry has devoted significant resources to production, marketing, and training with respect to standardized products (in economic terms, after the standard has become "locked in"), unexpected royalty demands from patent holders can have an extremely disruptive effect on the market. This phenomenon is commonly referred to as "hold-up," and results in an increase in the cost of standardized products, sometimes to levels that are inefficient and uncompetitive with alternative technologies.[101]

Patent ambush has been alleged in several cases within the ITC industry. The first of these cases to gain significant attention involved Dell Computer, which allegedly failed to disclose patents covering the voluntary VL-bus standard that it helped to develop at the Video Electronics Standards Association (VESA).[102] Following approval of the standard, Dell sought to enforce its patents against other computer manufacturers. Responding to complaints by these potential infringers, the FTC sought to enjoin Dell's enforcement activity on the ground that Dell's violation of the VESA disclosure rule effectively rendered its patents unenforceable. The FTC action resulted in the entry of a 1996 consent order that permanently prohibited Dell from enforcing its VL-bus patents against any third party.

Perhaps the most notorious case of disclosure failure within an SDO involved Rambus, Inc., a developer of semiconductor memory technology. Volumes have been written about the decade-long legal battles in which Rambus sought to assert its patents against implementers of dynamic random access memory (DRAM) technology standardized by the Joint Electron Device Engineering Council (JEDEC), a voluntary SDO in which Rambus participated in the early 1990s.[103] Despite conduct that was condemned by both the Court of Appeals for the Federal Circuit and the FTC, Rambus was largely exonerated on the basis that the poorly worded JEDEC patent policy was simply too imprecise an instrument to support liability. The court criticized the JEDEC policy as suffering from "a staggering lack of defining details" that left SDO participants with "vaguely defined expectations as to what they believe the policy requires."[104]

In a subsequent action, the FTC found Rambus liable for, among other things, attempted monopolization in violation of Section 2 of the Sherman Act and deceptive conduct under Section 5 of the FTC Act.[105] The FTC's decision, however, was reversed by the Court of Appeals for the D.C. Circuit, which held that Rambus's attempt to increase prices following adoption of a standard could not amount to anticompetitive conduct unless such conduct also resulted in adoption of the standard, which was not shown.[106] This decision has been criticized, both on the basis of its antitrust analysis and as a matter of public policy, inasmuch as it failed to sanction conduct that was widely condemned as deceptive.[107]

SDO Patent Policies

Policy Measures to Address Patent Issues

The Dell and Rambus cases demonstrated to the standards-development community both that patent ambush was a real possibility, and that existing SDO policies might lack adequate force to prevent such conduct. In response, many SDOs adopted or enhanced their formal policies to address ambush issues more clearly.[108] They did so by imposing one or both of the following obligations on SDO participants: (1) an obligation to disclose patents essential to implementation of a standard, and/or (2) an obligation to license patents essential to implementation of a standard, either on a royalty-free basis or on terms that are "reasonable and nondiscriminatory."[109] Such obligations are intended to ensure that standards developers have sufficient information to assess the relative costs and risks of technologies under consideration for standardization. That is, disclosure obligations ensure that standards developers know whether and which patents cover technologies under consideration, giving them the opportunity to "design around" such patents if they so wish.[110] Likewise, licensing obligations are intended to ensure that such patents will be licensed to implementers of the standard on terms that are, at least roughly, understood.[111]

Royalties for Standards-Essential Patents

Many SDO policies require that participants license patents essential to their standards on terms that are "reasonable and nondiscriminatory" (RAND) or "fair, reasonable, and nondiscriminatory" (F/RAND).[112] This requirement is built into ANSI's Essential Requirements for all ANSI-accredited SDOs[113] and is equally pervasive in Europe and other jurisdictions.[114] Despite the intuitive appeal of these requirements, however, a consistent and practical definition of F/RAND terms has proven difficult to develop.[115] In several recent cases, parties have disputed whether the terms under which licenses have been proffered violate or conform with F/RAND requirements.[116]

F/RAND licensing commitments tend to fail because there is no universal, objective standard by which reasonableness (or nondiscrimination) can be measured.[117] In order to make a F/RAND determination, the specific facts of each situation must be evaluated.[118] These facts include not only relevant market norms for royalties, but also customary practices relating to nonroyalty terms such as reciprocity, grant-back licenses, defensive suspension, confidentiality, and the like.[119] Also, given that a patent holder's F/RAND licensing terms are generally not revealed until negotiations that occur *after* a standard has been adopted (i.e., locked in), parties involved in standards setting can experience uncertainty regarding the ultimate cost of adopting a standard encumbered by patents. Put another way, the uncertainty of F/RAND licensing may simply substitute the risk of patent hold-up arising from unknown patents with hold-up arising from unknown F/RAND licensing terms.[120]

Ex Ante Disclosure of License Terms

Several commentators have suggested that permitting or requiring patent holders to disclose their royalty rates and licensing terms to SDO participants prior to the adoption of a standard (i.e., ex ante) would alleviate the F/RAND hold-up problems described in the previous section.[121] Such advance disclosure would enable SDO participants to evaluate the cost of including particular patented technologies in a standard prior to adoption, and would thus enable more efficient decision making with respect to the technical design of the standard.[122]

Critics, however, have argued that ex ante disclosures in the standards context might cause the standardization process to become more cumbersome, lengthy, and expensive.[123] However, a recent empirical study of the effects of ex ante disclosure of licensing terms on SDOs that have adopted ex ante disclosure policies found no evidence linking such policies to reductions in the number of standards adopted, the speed of standardization, the quality of standards, or the levels of royalties charged.[124] Concerns have also been raised that allowing ex ante licensing negotiations could facilitate the improper exchange of information among competitors and might place too much power in the hands of licensees acting collectively.[125] That is, potential implementers of a standard, in negotiating ex ante license terms with a patent holder, could collectively exert anticompetitive pressure on the patent holder, causing royalties to decrease below their fair (or optimal) level.[126] Following this argument to its logical conclusion, group pressure could drive all royalty rates toward zero, resulting in the devaluation of any patents covering a standard.

Despite these considerations, the U.S. Department of Justice (DOJ) has on two recent occasions issued Business Review Letters approving limited ex ante disclosure policies in SDOs. In the case of the VMEbus International Trade Association (VITA), the DOJ indicated in 2006 that it would not take enforcement action against an SDO that required participants to make ex ante declarations of the "most restrictive" licensing terms in their RAND licenses.[127] In approving the VITA policy, the DOJ concluded that ex ante disclosure of restrictive licensing terms would promote, rather than hinder, competition among patent holders. Likewise, in its 2007 Business Review Letter to the Institute for Electrical and Electronics Engineers (IEEE),[128] the DOJ approved a proposed arrangement in which patent holders were given the *option* to disclose their most restrictive licensing terms, including royalty rates, prior to the adoption of a standard. The DOJ called the IEEE proposal "a sensible effort to preserve competition between technological alternatives before the standard is set in order to alleviate concern that commitments by patent holders to license on RAND terms are not sufficient to avoid disputes." The European Commission has also expressed a general level of comfort with ex ante licensing disclosures in its guidelines relating to horizontal competition.[129]

BIOINFORMATICS STANDARDS POLICIES TODAY: ASSESSMENT AND RECOMMENDATIONS

Summary of Bioinformatics SDO Policies

Technical standards are likely to play an increasingly prominent role in the development, adoption, and regulation of bioinformatics research and technologies. Whether such standards relate to minimum information requirements, controlled vocabularies, or data exchange formats, the legal and policy considerations described in this section will play a key role in determining which of these standards are broadly adopted, and at what price.[130]

Given the importance of legal issues to standards development in other sectors of the economy, we conducted a review of the publicly available policies and rules of a number of major bioinformatics standards initiatives. The results, summarized in Table 10.1, indicate that few standardization efforts in bioinformatics address these issues in detail, and a significant number omit any legal guidelines in their public documentation.

Table 10.1
Summary of Selected Bioinformatics SDO Policies

Standards-Development Organization (SDO)	*Year Started*	*Notable Standards*	*Type*	*Antitrust Guidelines*	*Intellectual Property Policy*
Asia-Pacific Bioinformatics Network[131]	1998	Minimum Information About a Bioinformatics Investigation	Reporting	None	None
BioPAX.org[132]	2002	BioPAX	Exchange	None	Open source licenses that are academic and corporate friendly are used for all work created by the group.
European Bioinformatics Institute (EBI)[133]	1992	Systems Biology Ontology (SBO)	Terminology	None	No restrictions on the use or redistribution of data. Some original data may be subject to patent, copyright, or other intellectual property (IP) rights, and users must ensure that their exploitation of the data does not infringe the rights of such third parties.
Functional Genomics Data Society (FGED) [formerly Microarray Gene Expression Data (MGED) Society][134]	1999	MGED Ontology	Terminology	None	No funding sources have IP rights over FGED's output; FGED currently owns no patents or copyrights, and advocates free use of data, tools, and publications.
		Minimum Information About a Microarray Experiment (MIAME)	Reporting		
		Minimum Information Specification For *In Situ* Hybridization and Immunohistochemistry Experiments (MISFISHIE)	Reporting		
		MAGE-ML	Exchange		
Functional Genomics Investigation Ontology (FuGO) project[135]	2006	Functional Genomics Investigation Ontology (FuGO)	Terminology	None	None
Gene Ontology Consortium[136]	1998	Gene Ontology	Terminology	None	Unrestricted use is granted if user acknowledges GO as source, displays the version number, and does not alter files.
Genomic Standards Consortium[137]	2005	Minimum Information about a MARKer gene Sequence standard (MIMARKS)	Reporting	None	GSC utilizes Common Public License Version 1.0 (CPL). The license generally provides that any contributor grants a royalty-free license to use any copyrighted or patented materials that he or she contributes.
International Society Biocuration[138]	N/A	BioDBCore	Reporting	None	None

Table 10.1 (*continued*)
Summary of Selected Bioinformatics SDO Policies

Standards-Development Organization (SDO)	*Year Started*	*Notable Standards*	*Type*	*Antitrust Guidelines*	*Intellectual Property Policy*
Metabolomics Standards Initiative (MSI)[139]	2005	Core Information for Metabolomics Reporting (CIMR)	Reporting	None	Working group output is fully available to public.
OBO Foundry[140]	2001	Does not create its own, but endorses ontologies that meet its principles and standards	Terminology	None	OBO will not endorse ontology unless it allows for free use to all.
Proteomics Standards Initiative (PSI) [HUPO][141]	2002	Minimum Information about a Proteomics Experiment (MIAPE)	Reporting	None	PSI takes no position regarding validity or scope of any IP or other rights pertaining to the implementation or use of technology or the extent to which any license under such rights might or might not be available; neither does it represent that it has made any effort to identify any such rights.
		Minimum Information about a Molecular Interaction Experiment (MIMIx)	Reporting		
		Minimum Information about a Protein Affinity Reagent (MIAPAR)	Reporting		
		Proteomics Standards Initiative-Molecular Interaction (PSI-MI) XML	Exchange		
SBML.org[142]	2000	Systems Biology Markup Language	Exchange	None	The organization considers SBML to be free and open to the community, such that no one owns rights to it.
University of Auckland[143]	2004[144]	CellML	Exchange	None	Individuals may freely use, publish, and redistribute CellML; write and sell applications that create, load, or write CellML-valid XML files; distribute or sell their own CellML-valid XML files; and transmit verbatim copies of the CellML format to any person without restriction.
Worldwide Web Consortium (W3C)[145]	1994	Bio2RDF	Exchange	None	Bio2RDF is released under GPL v. 2.0; W3C will not approve a recommendation if it is aware that essential patent claims are not available on royalty-free terms.

As shown in Table 10.1, in many cases the organizations responsible for bioinformatics standards development either lack written policies entirely or have adopted vague, aspirational statements regarding a desire that materials produced be "open" and publicly available. This informal and minimalist approach not only invites opportunistic behavior but also leaves aggrieved participants with little legal recourse after abuse has occurred.

Nevertheless, the data collected in Table 10.1 does reveal a number of recurring themes. These include a general preference that bioinformatics standards be "open" and "not restricted" by intellectual property rights. Stated more precisely, these preferences can be expressed in terms of the following two maxims: (1) bioinformatics standards should be freely available, and (2) the implementation of bioinformatics standards should not be impeded by patent holders.

Policy Recommendations

As discussed above, it is important for SDOs to establish clear policies regarding compliance with antitrust law and the disposition of intellectual property rights in a manner that is aligned with the expectations and wishes of their participants. At a minimum, such policies should include provisions addressing the antitrust considerations described above under the heading "Policies and Precautions" in the section titled "Competition and Antitrust Issues for Standards Developers," as well as minimum provisions relating to copyrights and patents. Among the more critical issues to include in such policies are the following:

Copyrights

- What rights do SDO participants grant to the SDO and to one another with respect to their copyrights in documents, text, and code that have been "contributed" to a standard?
- Is the standard available free of charge, or must those wishing to implement it purchase copies from the SDO or others? If the latter, do SDO participants gain access at a discounted rate?
- What rights does the SDO grant to the world under its copyrights, if any, in a standard (e.g., can the standard be reproduced, modified, translated, excerpted, or otherwise manipulated outside of the SDO that developed it)?

Patents

- May SDO participants obtain patents covering the implementation of the SDO's standards?
- If so, must these participants agree to license such patents to all implementers of the standards?
- If so, must these licenses be royalty-free, or may royalties be charged (and if so, are there any limitations on royalty rates, F/RAND or otherwise)?
- Must participants disclose these patents and, if so, when?

While at first blush it might appear that the above issues can be addressed in a relatively straightforward and concise manner, SDO antitrust and intellectual property policies have become increasingly complex and lengthy over the years. This trend can generally be attributed to the desire of SDO participants, and their legal counsel, to anticipate a growing range of legal contingencies and to accommodate the diverse interests and business models of those participants. For example, in a typical SDO that focuses on the development of next-generation wireless telephony standards, participants might include handset manufacturers, chip designers, service providers (carriers), software developers, and businesses that do no more than acquire and assert patents—not to mention academics, consumer advocates, and government agencies. It is not surprising that developing a policy that accommodates the requirements of each of these groups can be difficult, time-consuming, and expensive, and can result in a complex policy that reflects myriad compromises required to appease these differing interests and business models.[146]

In contrast to the world of mobile telephony, participants in bioinformatics standards-development activities currently appear to consist primarily of academics, industrial scientists, and a handful of product vendors. This observation, coupled with the general consensus around the simple intellectual property-related maxims described earlier in this section under the heading "Summary of Bioinformatics SDO Policies," suggests that bioinformatics SDOs might be able to adopt legally prudent policies while avoiding the complexity, effort, and rancor often experienced in ICT-focused SDOs.

To this end, Appendix 1 contains a brief example of an SDO policy reflecting the two maxims noted above and other key provisions. While it is not intended that this exemplar be adopted verbatim by all bioinformatics SDOs, it is hoped that it might be useful as a starting point for consideration, and that it might provide scaffolding atop which an SDO may build more specialized provisions to address the particularized needs of its participants.[147]

Making Policies Binding

Even the most artfully drafted SDO policy is of little value unless it binds its participants and possibly other implementers and users of its standards. Simply announcing the SDO's policy on the SDO website, without affirmative assent by those parties that the SDO intends to bind, will make the policy more difficult to enforce.[148] Thus, SDOs have adopted a variety of approaches to ensure (or at least improve the chances) that their policies will legally bind their participants. These approaches include the incorporation of such policies into the SDO's bylaws and other constitutional documents, the use of signed participation agreements, and the issuance of frequent reminders of the existence and effect of such policies.[149]

Many bioinformatics standardization efforts are carried out in semi-academic settings without formal participation or membership structures. Groups of this nature would do well to consider formalizing their participation at least to the extent of requiring that participants expressly acknowledge (or, better still, agree to comply with) the SDO's written policies.

CONCLUSION

Though bioinformatics standards-development organizations have demonstrated a consistent preference for standards that are open and unencumbered by intellectual property rights, few of them have adopted explicit policies ensuring that these preferences will be respected by their participants. As evidenced by the increasing number of disputes and litigation in other fields that are heavily dependent on interoperability standards, it is important for organizations to adopt clear and binding policies governing the conduct of their participants. Accordingly, bioinformatics standards-development organizations should carefully review their existing policies and procedures. To the extent that they fail to address key points regarding process and intellectual property, these policies and procedures should be revised to reflect the intentions and expectations of participants in the standardization process.

NOTES

1. Am. Bar Ass'n, ABA Section of Antitrust Law, Handbook on the Antitrust Aspects of Standards Setting 1 (2d ed. 2011) [hereinafter ABA Antitrust Handbook].
2. *Id.* at 9.
3. Institute for Electrical and Electronics Engineers, http://www.ieee.org/about/index.html (last visited Jan. 30, 2012).
4. W3C, http://www.w3.org/ (last visited Jan. 30, 2012).
5. Asia-Pacific Bioinformatics Network, http://www.apbionet.org/ (last visited Jan. 30, 2012).
6. The European Bioinformatics Institute, http://www.ebi.ac.uk/ (last visited Jan. 30, 2012).
7. HUPO Proteomics Standards Initiative, http://www.psidev.info/ (last visited Jan. 30, 2012).
8. The Ontology for Biomedical Investigations, http://obi-ontology.org/page/Main_Page (last visited Jan. 30, 2012).
9. For a more detailed overview of the standards-development landscape, see Brad Biddle et al., *The Expanding Role and Importance of Standards in the Information and Communications Technology Industry*, 52 Jurimetrics 177 (2012); Am. Bar. Ass'n,, Committee on Technical Standardization, Section of Science & Technology Law, Standards Development Patent Policy Manual x–xi (Jorge L. Contreras ed., 2007) [hereinafter ABA Patent Policy Manual].
10. *See* ASTM Int'l, www.astm.org (last visited October 28, 2011).
11. Am. Nat'l Standards Inst., *ANSI Essential Requirements: Due Process Requirements for American National Standards* (2010); *available at* http://publicaa.ansi.org/sites/apdl/Documents/Standards%20Activities/American%20National%20Standards/Procedures,%20Guides,%20and%20Forms/2010%20ANSI%20Essential%20Requirements%20and%20Related/2010%20ANSI%20Essential%20Requirements.pdf.
12. *Standards Activities Overview*, American Nat'l Standards Inst., http://www.ansi.org/standards_activities/overview/overview.aspx?menuid=3 (last visited Jan. 21, 2012).
13. *Id.*
14. Int'l Org. for Standardization, www.iso.org (last visited Oct. 28, 2011).
15. Memorandum from the U.S. Office of Mgmt. & Budget to the Heads of Executive Dep'ts & Agencies, Circular No. A-119 Revised (Feb. 10, 1998), http://www.whitehouse.gov/omb/circulars_a119 [hereinafter OMB Circular A-119].
16. Final Act Embodying the Results of the Uruguay Round of Multilateral Trade Negotiations, Apr. 15, 1994, Agreement on Technical Barriers to Trade, Art. 2.4, 33 I.L.M. 1125 (1994), *available at* www.wto.org/english/docs_e/legal_e/03-fa.pdf.
17. *A Science Primer*, National Ctr. for Biotechnology Info. (2004), http://www.ncbi.nlm.nih.gov/About/primer/bioinformatics.html (last visited Jan. 30, 2012).
18. This classification system was developed by Biosharing.org, which catalogs and provides information regarding most of the standards discussed below. *See A Catalogue of Standards*, Biosharing, http://www.biosharing.org/standards (last visited October 28, 2011). A list of the standards compiled by Biosharing.org can be found at http://www.biosharing.org/standards_view.
19. Thomas R. Gruber, *Toward Principles for the Design of Ontologies Used for Knowledge Sharing?*, 43 Int'l J. Hum.-Computer Stud. 907, 908 (1995).
20. *See* Michael Ashburner et al., *Gene Ontology: Tool for the Unification of Biology*, 25 Nature Genetics 25 (2000).
21. *See* Gene Ontology Consortium, *The Gene Ontology (GO) Project in 2006*, 34 Nucleic Acid Research D322, D323 (2006).
22. The Gene Ontology Consortium, *The Gene Ontology in 2010: Extensions and Refinements*, 38 Nucleic Acid Research D331, D332 (2009).
23. *See Systems Biology Ontology*, European Molecular Biology Lab.-European Bioinformatics Inst., http://www.ebi.ac.uk/sbo/main (last visited Oct. 28, 2011).
24. *See* Barry Smith, *The OBO Foundry: Coordinated Evolution of Ontologies to Support Biomedical Data Integration*, 25 Nature Biotechnology 1251 (2007).
25. *See* Patricia L. Whetzel et al., *Development of FuGO: An Ontology for Functional Genomics Investigations*, 10 OMICS 199 (2006).

26. *See* Catherine A. Ball & Alvis Brazma, *MGED Standards: Work in Progress*, 10 OMICS 138, 139–40 (2006).

27. There are more comprehensive lists of minimum information standards projects available. Taylor, *supra* note 22, at 528; Lyle D. Burgoon, *Clearing the Standards Landscape: The Semantics of Terminology and Their Impact on Toxicologenomics*, 99 Toxicological Sci. 403, 408–10 (2007).

28. *See* Inst. for Sys. Biology, MISFISHIE, http://scgap.systemsbiology.net/standards/misfishie/ (last visited Oct. 28, 2011).

29. *See* Sandra Orchard & Henning Hermjakob, *The HUPO Proteomics Standards Initiative—Easing Communication and Minimizing Data Loss in a Changing World*, 9 Briefings in Bioinformatics 166, 167 (2008).

30. *See id.* at 170–71.

31. *See* Pelin Yilmaz, *The Genomic Standards Consortium: Bringing Standards to Life for Microbial Ecology*, 5 ISME J. 1565 (2011).

32. *See* Chris F. Taylor, *Promoting Coherent Minimum Reporting Guidelines for Biological and Biomedical Investigations: The MIBBI Project*, 26 Nature Biotechnology 889, 889 (2008).

33. Michael Y. Galperin & Guy R. Cochrane, *The 2011 Nucleic Acids Research Database Issue and the Online Molecular Biology Database Collection*, 39 Nucleic Acids Res. D1, D1 (2011).

34. James H. Coombs et al., *Markup Systems and the Future of Scholarly Text Processing*, 30 Comm. ACM 933 (1987).

35. *Introduction to HTML 4*, W3C, http://www.w3.org/TR/html4/intro/intro.html (last visited Jan. 30, 2012).

36. *Extensible Markup Language Introduction*, W3C, http://www.w3.org/XML/ (last visited Jan. 30, 2012).

37. More comprehensive lists of bioinformatics markup languages can be found in Burgoon, *supra* note 27, at 409; Luciano Milanesi, *Trends in Modeling Biomedical Complex Systems*, 10 BMC Bioinformatics 12 (Supp. 2009).

38. *See* M. Hucka, *The Systems Biology Markup Language (SBML): A Medium for Representation and Exchange of Biochemical Network Models*, 19 Bioinformatics 524 (2003).

39. *See* Catherine M. Lloyd et al., *CellML: Its Future, Present and Past*, 85 Progress in Biophysics & Molecular Biology 433, 436 (2004).

40. *See* Ball & Brazma, *supra* note 26, at 140–41.

41. *See* Christoph Wierling et al., *Resources, Standards and Tools for Systems Biology*, 6 Briefings Functional Genomics & Proteomics 240, 245–46 (2007).

42. *See, e.g.*, John Quackenbush, *Standardizing the Standards*, 2 Molecular Systems Biology 1, 1–2 (2006); Xiaoshu Wang et al., *From XML to RDF: How Semantic Web Technologies Will Change the Design of 'omic' Standards*, 23 Nature Biotechnology 1099 (2005).

43. Wang, *supra* note 42, at 1099.

44. *See* Francois Bellau et al., *Bio2RDF: Towards a Mashup to Build Bioinformatics Knowledge Systems*, 41 J. Biomedical Informatics 706 (2008).

45. Philippe LeHegaret et al., *What Is Document Object Model?*, W3C, Apr. 7, 2004, http://www.w3.org/TR/DOM-Level-3-Core/introduction.html (last visited Feb. 1, 2012).

46. *MicroArray Gene Experiment—Object Model*, Functional Genomics Data Society, http://www.mged.org/Workgroups/MAGE/mage.html (last visited Oct. 29, 2011).

47. Burgoon, *supra* note 27, at 409.

48. *Functional Genomics Experiment Object Model*, Functional Genomics Experiment, http://fuge.sourceforge.net/index.php (last visited Oct. 29, 2011).

49. *See* Pascale Gaudet et al., *Towards BioDBCore: A Community-Defined Information Specification for Biological Databases*, 39 Nucleic Acids Res. D7 (2011).

50. *Id.*

51. 15 U.S.C. § 1 (2006).

52. 15 U.S.C. § 45 (2006).

53. *See generally* ABA Antitrust Handbook, *supra* note 1, at 51–52.

54. *Id.*

55. *Id.*

56. Nat'l Ass'n of Review Appraisers & Mortg. Underwriters v. Appraisal Found., 64 F.3d 1130, 1133 (8th Cir. 1995) (upholding Appraisal Foundation's membership criteria where there was no evidence of any harm to the overall market for appraisal organizations).

57. *See, e.g.*, Pretz v. Holstein Friesian Ass'n of Am., 698 F. Supp. 1531, 1542 (D. Kan. 1988) (holding that cattle registry association did not commit antitrust violation by expelling a member alleged to be manipulating tests for grading cattle).

58. 152 F.3d 48 (1st Cir. 1998).

59. *Id.* at 53; *see also* True Position, Inc. v. LM Ericsson Tel. Co., No. 2011-CV-04574-RK, 2012 WL 33075 (E.D. Pa. Jan. 6, 2012) (alleging that concerted activity by defendants, including two SDOs, resulted in exclusion of plaintiff's technology from wireless telecommunications standard); Golden Bridge Tech. v. Nokia Inc., No. 6:06-CV-163, 2007 U.S. Dist. LEXIS 67071, at *25–26 (E.D. Tex. Sept. 10, 2007) (plaintiff unsuccessfully claimed that defendants intentionally excluded its technology from standard relating to wireless telephony).

60. *See, e.g.*, United States v. Container Corp. of Am., 393 U.S. 333, 337 (1969) (holding that exchange of price quotes for sales of corrugated containers was a price-fixing agreement that violated the Sherman Act since the product was fungible and demand was inelastic); Todd v. Exxon Corp., 275 F.3d 191, 214–15 (2d Cir. 2001) (holding that employees have a claim against employers to examine employers' habit of sharing information regarding nonmanagerial, nonunion workers to determine whether it was a collusion to keep wages down in violation of the Sherman Act).

61. *See* Harry S. Gerla, *Federal Antitrust Law and Trade and Professional Association Standards and Certification*, 19 Dayton L. Rev. 471, 486 (1994).

62. *See, e.g.*, United States v. Nat'l Ass'n of Realtors, No. 05 C 5140, 2006 U.S. Dist. LEXIS 86963, at *48 (N.D. Ill. Nov. 27, 2006).

63. 456 U.S. 556 (1982).

64. *Id.* at 577.

65. Consol. Metal Prod., Inc. v. Am. Petroleum Inst., 846 F.2d 284 (5th Cir. 1988).

66. *Id.* at 293; *see generally* ABA Antitrust Handbook, *supra* note 1, at 55–59.

67. 486 U.S. 492 (1988).

68. *See, e.g.*, *Promoting Competition and Innovation: What You Need to Know About the IEEE Standards Association's Antitrust and Competition Policy*, IEEE Standards Ass'n, Aug. 24, 2010, http://standards.ieee.org/develop/policies/antitrust.pdf; ABA Antitrust Handbook, *supra* note 1, at 69–70.

69. Mark A. Lemley, *Intellectual Property Rights and Standard-Setting Organizations*, 90 Cal. L. Rev. 1889, 1896 (2002).

70. *See, e.g.*, Int'l Org. for Standardization and Int'l Electrotechnical Comm'n, Copyright, Standards and the Internet (2007), *available at* http://www.iso.org/iso/copyright_information_brochure.pdf.

71. *Id.*

72. *See, e.g.*, *Legal Provisions Relating to IETF Documents*, IETF Trust, § 3, Dec. 28, 2009, http://trustee.ietf.org license-info/archive/IETF-Trust-License-Policy-20091228.pdf.

73. *See* Pamela Samuelson, *Questioning Copyrights in Standards*, 47 B.C. L. Rev. 193, 197 (2007).

74. HCFA was renamed the Centers for Medicare & Medicaid Services on July 1, 2001. *See* Centers for Medicare & Medicaid Services, http://www.cms.gov/.

75. *About CPT*, Am. Med. Ass'n, http://www.ama-assn.org/ama/pub/physician-resources/solutions-managing-your-practice/coding-billing-insurance/cpt/about-cpt.shtml (last visited Nov. 6, 2011).

76. *See* Samuelson, *supra* note 73, at 197; Dale R. Kurth, *Contracting Around Copyright? An Introduction to Copyright Misuse*, Landslide, Jan./Feb. 2010, at 44, 47.

77. Practice Mgmt. Info. Corp. v. Am. Med. Ass'n, 121 F.3d 516, 517 (9th Cir. 1997), *opinion amended by* 133 F.3d 1140 (9th Cir. 1998).

78. *Id.* at 518.

79. *Id.*; *see* Samuelson, *supra* note 73, at 197.

80. *Practice Mgmt. Info. Corp.*, 121 F.3d at 521.

81. *See* OMB Circular A-119, *supra* note 15, § 6(j).

82. 293 F.3d 791 (5th Cir. 2002) (en banc), *cert. denied*, 537 U.S. 1043 (2002).

83. *Id.* at 793.

84. *Id.*

85. *Id.* at 794.

86. *Id.* at 800.

87. *Id.*; *see also* Practice Mgmt. Info. Corp. v. Am. Med. Ass'n, 121 F.3d 516, 516 (9th Cir. 1997), *opinion amended by* 133 F.3d 1140 (9th Cir. 1998); CCC Info. Serv. v. Maclean Hunter Mkt. Reports, Inc., 44 F.3d 61 (2d Cir. 1994).

88. Letter from Peter L. Strauss to Office of the Federal Register dated February 10, 2012 (http://www.americanbar.org/content/dam/aba/administrative/administrative_law/2012_feb_4-5_council_agenda_with_materials.authcheckdam.pdf). The petition largely reflects a December 2011 recommendation by the Administrative Conference of the United States (ACUS). Admin. Conf. of the U.S., Administrative Conference Recommendation 2011-5: Incorporation by Reference (Adopted Dec. 8, 2011), *available at* http://www.americanbar.org/content/dam/aba/administrative/administrative_law/2012_feb_4-5_council_agenda_with_materials.authcheckdam.pdf.

89. *See, e.g.*, Letter from Michael Hertz to Hon. Cass Sunstein, Administrator, Office of Information and Regulatory Affairs dated June 1, 2012 (submitted on behalf of the ABA Section of Administrative Law).

90. *See, e.g.*, Am. Nat'l Standards. Inst., ANSI Response to Request for Comments on Incorporation by Reference (Apr. 10, 2012), *available at* http://www.x12.org/docs/ANSI%20Response%20IBR_041012.pdf.

91. The economic and legal literature exploring this phenomenon, both within and outside the context of technical standards, is extensive and varied. *See, e.g.*, Einer Elhauge, *Do Patent Holdup and Royalty Stacking Lead to Systematically Excessive Royalties?*, 4 J. Competition L. & Econ. 535 (2008); Mark Lemley & Carl Shapiro, *Patent Holdup and Royalty Stacking*, 85 Tex. L. Rev. 1991 (2007); Carl Shapiro, *Navigating the Patent Thicket: Cross-Licenses, Patent Pools and Standard Setting*, 1 Innovation Pol'y & Econ. 119 (Adam B. Jaffe et al. eds., 2004).

92. Lemley & Shapiro, *supra* note 91, at 1993.

93. U.S. Dep't of Justice & U.S. Fed. Trade Comm'n, Antitrust Enforcement and Intellectual Property Rights: Promoting Innovation and Competition 76–84 (2007), *available at* http://www.ftc.gov/reports/innovation/P040101 PromotingInnovationandCompetitionrpt0704.pdf [hereinafter 2007 DOJ/FTC Report].

94. Letter from Joel I. Klein, Assistant Attorney Gen., U.S. Dep't of Justice, to Garrard R. Beeney (June 26, 1997), *available at* http://www.justice.gov/atr/public/busreview/215742.htm.

95. Letter from Joel I. Klein, Assistant Attorney Gen., U.S. Dep't of Justice, to Garrard R. Beeney (Dec. 16, 1998), *available at* http://www.justice.gov/atr/public/busreview/2121.htm; Letter from Joel I. Klein, Assistant Attorney Gen., U.S. Dep't of Justice, to Carey R. Ramos (June 10, 1999), *available at* http://www.justice.gov/atr/public/busreview/2485 .htm.

96. Letter from Charles A. James, Assistant Attorney Gen., U.S. Dep't of Justice, to Ky P. Ewing (Nov. 12, 2002), *available at* http://www.justice.gov/atr/public/busreview/200455.htm.

97. 2007 DOJ/FTC Report, *supra* note 93, at 68–71.

98. Vizio, Inc. v. Funai Elec. Co. Ltd., 24 F.C.C. Rcd. 2880 (2009).

99. Vizio, Inc. v. Int'l Trade Comm'n, 605 F.3d 1330 (Fed. Cir. 2010).

100. *See* U.S. Federal Trade Commission, The Evolving IP Marketplace: Aligning Patent Notice and Remedies with Competition (2011) [hereinafter FTC 2011 Report]; Elhauge, *supra* note 91; Mark A. Lemley, *Ten Things to Do About Patent Holdup of Standards (and One Not to)*, 48 B.C. L. Rev., 149 (2007); Sean M. Royall, et al., *Deterring "Patent Ambush" in Standard Setting: Lessons Learned from* Rambus *and* Qualcomm, Antitrust, Summer 2009, at 34.

101. *See* Lemley, *supra* note 100, at 154–55 and FTC 2011 Report, *supra* note 100, at 5:

> [a]t the time a manufacturer faces an infringement allegation, switching to an alternative technology may be very expensive if it has sunk costs in production using the patented technology. . . . If so, the patentee can use the threat of an injunction to obtain royalties covering not only the market value of the patented invention, but also a portion of the costs that the infringer would incur if it were enjoined and had to switch. This higher royalty based on switching costs is called the "hold-up" value of the patent. Patent hold-up can over-compensate patentees, raise prices to consumers who lose the benefits of competition among technologies, and deter innovation by manufacturers facing the risk of hold-up.

102. *In re* Dell Computer Corp., 121 F.T.C. 616 (1996).

103. *See, e.g.*, ABA Patent Policy Manual, *supra* note 9, at vii–viii; Lemley, *supra* note 100, at 1929–30; Joel M. Wallace, Rambus v. F.T.C. *in the Context of Standard-Setting Organizations, Antitrust, and the Patent Hold-Up Problem*, 24 Berkeley Tech. L.J. 661 (2009); Robert A. Skitol & Kenneth M. Vorrasi, *Patent Holdup in Standards Development: Life After* Rambus v. FTC, Antitrust, Summer 2009, at 26.

104. Rambus, Inc. v. Infineon Techs. AG, 318 F.3d 1081, 1102 (Fed. Cir. 2003), *cert. denied,* 540 U.S. 874 (2003).

105. Opinion of the Commission, *In re* Rambus, Inc., Docket No. 9302, slip op. at 3 (F.T.C. Aug. 2, 2006) (Public Record Version), *available at* http://www.ftc.gov/os/adjpro/d9302/060802commissionopinion.pdf.

106. Rambus, Inc. v. F.T.C., 522 F.3d 456, 463 (D.C. Cir. 2008), *cert. denied,* 129 S. Ct. 1318 (2009).

107. *See, e.g.,* Wilson D. Mudge & Marni B. Karlin, *The District of Columbia Circuit Court's Decision in Rambus Overturns the FTC Finding of Monopolization*, Intel. Prop. & Tech. L.J. 11 (2008); Wallace, *supra* note 103, at 683–90.

108. Patent ambush can involve either patents held by participants in the SDO or patents held by nonparticipating third parties. The risk posed by SDO participants' patents is perceived as particularly serious, however, because, unlike nonparticipating third parties, SDO participants can potentially shape the technical parameters of a standard toward their own patent positions. *See* Lemley, *supra* note 100, at 52; Royall, *supra* note 100, at 34.

109. *See* 2007 DOJ/FTC Report, *supra* note 93, at 42–48; ABA Patent Policy Manual, *supra* note 9, at xiii–xiv; Joseph Farrell et al., *Standard Setting, Patents and Hold-Up*, 74 Antitrust L.J. 603, 624–43 (2007).

110. ABA Patent Policy Manual, *supra* note 9, at xiv.

111. *Id.*

112. *See id.* at 56–58.

113. Am. Nat'l Standards Inst., *Guidelines for Implementation of the ANSI Patent Policy* (1997), *available at* http://publicaa.ansi.org/sites/apdl/Documents/Standards%20Activities/American%20National%20Standards/Procedures,%20 Guides,%20and%20Forms/Guidelines%20for%20Implementation%20of%20ANSI%20Patent%20Policy%202011.pdf.

114. *See, e.g.*, Anne Layne-Farrar et al., *Pricing Patents for Licensing in Standard-Setting Organizations: Making Sense of FRAND Commitments*, 74 Antitrust L.J. 671 (2007).

115. ABA Antitrust Handbook, *supra* note 1, at 216.

116. *See, e.g.*, Jorge L. Contreras, *The FRAND Wars: Who's on First?*, Patently-O (Apr. 17, 2012), *available at* http://www.patentlyo.com/patent/2012/04/the-frand-wars-whos-on-first.html.

117. *See, e.g.*, Anne Layne-Farrar et al., *Pricing Patents for Licensing in Standard-Setting Organizations: Making Sense of FRAND Commitments*, 74 Antitrust L.J. 671 (2007).

118. Wallace, *supra* note 103, at 664.

119. *Id.* at 665.

120. *See id.*; Justin Hurwitz, *The Value of Patents in Industry Standards: Avoiding License Arbitrage with Voluntary Rules*, 36 A.I.P.L.A. Q.J. 3, 3–4 (2008). *But see* Joseph Scott Miller, *Standard Setting, Patents, and Access Lock-In: RAND Licensing and the Theory of the Firm*, 40 Ind. L. Rev. 351, 357 (2007) (arguing that RAND obligations are not "materially underspecified").

121. Lemley, *supra* note 100, at 158–59; Gil Ohana et al., *Disclosure and Negotiation of Licensing Terms Prior to Adoption of Industry Standards: Preventing Another Patent Ambush?*, 24 Eur. Competition L. Rev., 644, 648–50 (2003); Robert A. Skitol, *Concerted Buying Power: Its Potential for Addressing the Patent Holdup Problem in Standard Setting*, 72 Antitrust 727, 741–42 (2005); Wallace, *supra* note 103, at 689–92.

122. Wallace, *supra* note 103, at 690.

123. 2007 DOJ/FTC Report, *supra* note 93, at 50; Skitol, *supra* note 121, at 734.

124. Jorge L. Contreras, *Nat'l Inst. Standards & Tech. Study Report—An Empirical Study of the Effects of Ex Ante Licensing Disclosure Policies on the Development of Voluntary Technical Standards* (GCR 11-934, June 27, 2011), *available at* http://gsi.nist.gov/global/docs/pubs/NISTGCR_11_934.pdf.

125. 2007 DOJ/FTC Report, *supra* note 93, at 52.

126. *See* 2007 DOJ/FTC Report, *supra* note 93, at 52–53; Skitol, *supra* note 121, at 735.

127. Letter from Thomas O. Barnett, Assistant Attorney General, U.S. Dep't of Justice, to Robert A. Skitol on behalf of VMEbus International Trade Association (Oct. 30, 2006), *available at* http://www.justice.gov/atr/public/busreview/219380.htm.

128. Letter from Thomas O. Barnett, Assistant Attorney General, U.S. Dep't of Justice, to Michael A. Lindsay on behalf of Institute of Electrical and Electronics Engineers (Apr. 30, 2007), *available at* http://www.justice.gov/atr/public/busreview/222978.htm.

129. European Commission, Guidelines on the Applicability of Article 101 of the Treaty on the Functioning of the European Union to Horizontal Co-operation Agreements, ¶ 299 (2011).

130. Admittedly, intellectual property considerations may play a smaller role in standards devoted to the definition of controlled vocabularies and minimum information guidelines (which are themselves less amenable to patent protection) than they do in standards relating to data exchange formats, database structures, and scientific apparatus. This distinction may be less pronounced with respect to the impact and importance of antitrust issues.

131. Asia-Pacific Bioinformatics Network, http://www.apbionet.org (last visited Jan. 30, 2012).

132. BioPAX.org, http://www.biopax.org (last visited Jan. 30, 2012).

133. The European Bioinformatics Inst., http://www.ebi.ac.uk (last visited Jan. 30, 2012).

135. Functional Genomics Data Society, http://www.mged.org (last visited Jan. 30, 2012).

134. The Ontology for Biomedical Investigations, http://obi-ontology.org/page/Main_Page (last visited Jan. 30, 2012).

138. Gene Ontology Consortium, http://www.geneontology.org/GO.consortiumlist.shtml (last visited Jan. 30, 2012).

136. Genomic Standards Consortium, http://gensc.org/gc_wiki/index.php/Main_Page (last visited Jan. 30, 2012).

137. Int'l Society of Biocuration, http://biocurator.org (last visited Jan. 30, 2012).

139. Metabolomics Standards Initiative, http://msi-workgroups.sourceforge.net (last visited Jan. 30, 2012).

140. The Open Biological and Biomedical Ontologies, http://obofoundry.org (last visited Jan. 30, 2012).

141. Proteomics Standards Inst., http://www.psidev.info (last visited Jan. 30, 2012).

142. The Systems Biology Markup Language, http://sbml.org/Main_Page (last visited Jan. 30, 2012).

143. University of Auckland, http://www.abi.auckland.ac.nz/uoa/cellml (last visited Jan. 30, 2012).

144. Work on CellML began in 2004. The university itself is much older.

145. W3C, http://www.w3.org/ (last visited Jan. 30, 2012).

146. For example, the most recent Standards Board Bylaws of the IEEE Standards Association run to 18 pages. IEEE Standards Ass'n, IEEE-SA Standards Board Bylaws (Dec. 2011), *available at* http://standards.ieee.org/develop/policies/bylaws/sb_bylaws.pdf. In addition, the IEEE-SA maintains a separate Copyright Policy, Operations Manual, and Standards Board Operations Manual.

147. The ABA Patent Policy Manual, *supra* note 9, contains a large collection of annotated sample clauses that appear in SDO patent policies.

148. *See, e.g.*, Specht v. Netscape Commc'ns Corp., 306 F.3d 17 (2d Cir. 2002) (holding that a user of an online service is not deemed to have agreed to a vendor's terms of service available when those terms were not conspicuous and the user was not required to assent to them explicitly).

149. *See* ABA Patent Policy Manual, *supra* note 9, at 27.

APPENDIX 1

Sample Bioinformatics "Open" Standards Development Policy

[*Insert name of standards development organization* (SDO)] is committed to the open and transparent development and dissemination of technical standards relating to [*subject matter*]. Accordingly, all meetings, working groups, mailing lists, correspondence, and other activities conducted under the auspices of SDO ("SDO Activities") will be conducted in accordance with the following rules.

1. No Anticompetitive Conduct.

No SDO Participant* shall, during or in connection with any SDO Activity:

a. discuss product prices, profits, internal cost, bidding, terms of bidding, allocation of customers, division of sales markets, sales territories, or marketing strategies;
b. condition or discuss conditioning the implementation of an SDO standard on the implementer's use of products or services offered by a particular party; or
c. agree with others to refuse to do business with a particular party.

2. Copyright License.

a. Each SDO Participant hereby grants to SDO and each other SDO Participant a non-exclusive, perpetual, royalty-free copyright license to reproduce, modify, distribute, and create derivative works of every document, proposal, text, code, comment, or other written item submitted by such SDO Participant to SDO or other SDO Participants in connection with any SDO Activity that is intended to influence or be incorporated into any draft or final standard (a "Contribution"). SDO Participants otherwise retain all rights in their respective Contributions.
b. Each SDO Participant agrees that the copyright in any final standard published by SDO shall be owned by SDO, subject to SDO Participants' ownership of their individual Contributions.

*The term "SDO Participant" must be defined to include all members, organizers, promoters, and other participants in SDO Activities.

3. Non-Assertion of Patents.

a. Each SDO Participant agrees that it shall not, at any time during or after its participation in SDO, assert or seek to collect royalties on any patent that is necessary to implement an SDO standard ("Necessary Patent") against any product or process that implements that SDO standard.
b. Each SDO Participant shall ensure that the obligations in Paragraph 3.a bind the transferee of any Necessary Patent, and shall notify any actual or proposed transferee of any Necessary Patent of such obligations.

4. No Confidential Information.

No Contribution submitted by any SDO Participant shall be deemed to be proprietary or confidential, and neither SDO nor any SDO Participant shall have any obligation to maintain any Contribution as confidential.

CHAPTER 11

Database Protection

Jon M. Garon

"Bioinformatics involves the acquisition, organization, storage, analysis and visualization of information contained within biological molecules."[1] This chapter focuses on the legal issues associated specifically with the organization of these data, through the technological, contractual, and legal regimes under which these databases are populated and operated. Here we will define a "bioinformatics database" to be one constructed for the storage, dissemination, and analysis of biological data. A bioinformatics database will generally involve "persistent sets of data that are maintained in a consistent state over essentially indefinite periods of time."[2] To be of any use, all legal and practical methods of protecting the database must ensure the persistency—integrity, reliability, and resiliency—of the data over time.

The access to a bioinformatics database may be designed to be entirely open to anyone; limited to authorized researchers and partners; or entirely proprietary. Similarly, the ability to add information to the database can range from entirely open to entirely closed, with any number of permutations in between.

For many projects, a community-based approach is essential. "The discipline of molecular biology and bioinformatics relies greatly on the use of community knowledge, rather than laws and axioms, to further understanding, and knowledge generation."[3] In practice, the "biological information resources, such as community databases, must be capable at some level of working together, of interoperating, so that users may interact with them collectively as a federated information infrastructure."[4]

Among the key factors for a database regime to be successful is the interoperability of databases so that different data sets can be mapped to each other.[5] This in turn requires either a common database schema or the ability to map heterogeneous data sets to one another. The National Institutes of Health (NIH), for example, through the National Centers for Biomedical Computing emphasizes the common system known as "Biositemaps"[6] to allow for "(i) locating, (ii) querying, (iii) composing or combining, and (iv) mining biomedical resources."[7]

At the same time, the significant capital investment in bioinformatics projects and potential revenue in the industry drives other projects to seek a wholly proprietary closed environment. In such cases, there may be no need to utilize a common database schema. Between the "open" projects that are designed to provide interoperability and sharing of data in the public sphere and the "proprietary" projects that allow no access and incorporate no other data, there may also be "blended" projects that include both a public component and a proprietary component.

In the United States, the law protects the order selection and arrangement of data under copyright law,[8] database data using trade secret law,[9] and the schema or data analysis under patent law.[10] Under the European Database Directive, EU member states extend additional legal protections for the compilation of specified data sets as a sui generis right.[11] In most cases, however, the effectiveness of the legal regime is determined by the contractual terms used to define and incorporate legal rights into the database usage.

CONTRACT RIGHTS IN DATABASES

Despite operating in the shadow of intellectual property rights, in almost every situation, the real relationship between the parties will be mediated by contract.[12] The contracts may be presented in any number of formats, including as traditional written agreements; as the terms and conditions of grant funding; or as the terms and conditions of access or interoperability among computer systems. If the contracting is done through software or computers, the threshold approach for such contracts should be through mandatory user interfaces that require an opt-in assent, commonly known as the "clickwrap" agreement.

Historically, contract law could not supplant property law or intellectual property law, because each person had to agree to assent to the contract before it could be enforceable. The computer age has changed this because it is possible that every person attempting to gain access to data on a computer server or through a website can be forced to enter through a particular digital doorway. Like a gated park, each entrant can be required to agree to the terms of the agreement as a condition of entry.

A properly executed clickwrap agreement has the following attributes: The contract terms are clearly readable in a scrollable window or linked page; the terms can be printed and downloaded; the transaction requires a box to check or radio button on which to click to demonstrate knowing assent; and that affirmative click is labeled with an explicit statement that the transaction or agreement is subject to the terms and conditions that are provided.[13] Less stringent notice of assent to online computer-mediated agreements has been found enforceable,[14] but there simply is no reason to design access to a database without each of these features when dealing with scientific information.

Generally speaking, clickwrap agreements have been upheld as enforceable. Like other contracts, the terms must not violate public policy or be substantively unconscionable.[15] This should not undermine the enforceability of agreements regarding reasonable terms and conditions for the use of bioinformatics data. Instead, such restrictions typically focus on onerous choice-of-venue provisions and provisions relating to dispute resolution clauses.[16]

Importantly, the contract conditioning the terms and conditions on the use of the data is independent of any legal property right in the data.[17] So it does not matter to the enforceability of the agreement that patent or copyright laws cannot protect the data.

For a community-based project, a master-use agreement can be drafted that incorporates the essential community understandings of the data's open access, any obligations of attribution, and even any downstream obligations prohibiting—or permitting—proprietary projects based upon the open-access data.[18] Like open source software, an open-access protocol could contractually require that published research be free to the public, if the participants to that community choose to do so. But such downstream restrictions must be explicit in the agreement and assented to by each participant through a fully transparent clickwrap or other contracting mechanism.

For proprietary projects, the duties of confidentiality, obligation to restrict access to the data, and protection of the security of the information must be included in the agreement. If the data includes identifiable patient data or personal health information, additional restrictions must be included in the agreement to comply with applicable federal regulations.[19]

For projects that provide access to alter the data or add additional data, additional contract provisions should specify the standards—both biomedical and technical—required for any researcher or data set to be incorporated into the common database. These standards may well be incorporated into the agreement by reference so that the technical protocols need only be identified. This will make the data standards become the governing law of the database.

If access to the database is not restricted by contract and software, then the standard rules of intellectual property apply. In the United States, they provide only modest legal protection.

U.S. COPYRIGHT PROTECTION FOR DATABASES

Order Selection and Arrangement

U.S. copyright law protects "original works of authorship fixed in any tangible medium of expression"[20] while excluding facts, ideas, and discoveries.[21] In *Feist Publications, Inc. v. Rural Telephone Service Co.*,[22] the Supreme Court addressed the extent to which the law could protect the effort that goes into collecting, correcting, and maintaining factual information.

The Supreme Court explained that the need for originality is a foundational, constitutional requirement of copyright law.[23] Flowing from this constitutional basis, the court explained the law's protection of facts and factual compilations. "It is this bedrock principle of copyright that mandates the law's seemingly disparate treatment of facts and factual compilations. 'No one may claim originality as to facts.'"[24] The Court illustrated its point using census data to explain that census takers do not author the census information but instead take the data from the world around them.[25] In precisely the same manner, biological data exists in nature and therefore cannot be protected by U.S. copyright law. The data themselves cannot be the subject of copyright.

Admittedly, at least a significant portion of the data is not mathematical but is instead narrative. "[M]uch bioinformatics data is kept in a natural language text-based form, in either annotations or bibliographic databases."[26] Nonetheless, short phrases are not sufficient to receive copyright protection, nor are narrative descriptions copyrightable when there are only a limited number of variations on the way to describe something.[27]

Known as the *merger* doctrine, "expression is not protected in those instances where there is only one or so few ways of expressing an idea that protection of the expression would effectively accord protection to the idea itself."[28] The narrative descriptors of bioinformatics data sets fall squarely within the types of facts that are precluded from being protected by copyright; any descriptions would have the effect of limiting access to the facts rather than merely the expression of those facts. "Facts, whether alone or as part of a compilation, are not original and therefore may not be copyrighted."[29]

At the same time, however, there is copyright protection available to some forms of factual compilations. "A factual compilation is eligible for copyright if it features an original selection or arrangement of facts, but the copyright is limited to the particular selection or arrangement. In no event may copyright extend to the facts themselves."[30] As defined in the Copyright Act, "[a] 'compilation' is a work formed by the collection and assembling of preexisting materials or of data that are selected, coordinated, or arranged in such a way that the resulting work as a whole constitutes an original work of authorship."[31] It is the creative selection, coordination, and arrangement of these data that sets it apart from other databases and constitutes creative work.

The selection, coordination, and arrangement must serve as the authorship for the database as a whole, and for many scientific studies, this may be a difficult standard to meet. In *Feist*, the subject of the dispute was a telephone white page directory. The choice to include all telephone subscribers was not authorial; it was merely mechanical. In the same way, any study of all genes in a sequence or all volunteers in a study may fail to provide the sufficient modicum of creativity necessary to create a copyright in the compilation of the data.

This problem is further exacerbated if the data are subject to rearrangement and reordering by the user. In such a case, there is no authorship in the default order of the data set—meaning that one of the attributes that could support authorship has been eliminated.

At the other end of the creativity spectrum, the creation of a new taxonomy or schema for data may be an appropriate subject matter for copyright. Such a new set of parameters and original organizing methodology may meet the copyright test if it is sufficiently expressive.[32] Copyright in this system could not be used to block access to the underlying data but could certainly be part of a broader system to render the data proprietary.

U.S. TRADE SECRET PROTECTION FOR VALUABLE, NONPUBLIC INFORMATION

Although the underlying data cannot be protected by copyright, those data that meet the criteria for trade secret protection may be afforded legal protection. Trade secret law is primarily a creature of state law, with each state having laws to protect from the theft of confidential, valuable, protected information.[33] There is also a federal criminal trade secret statute.[34] According to the Uniform Trade Secrets Act (UTSA),

> "Trade secret" means information, including formula, pattern, compilation, program, device, method, technique, or process, that: (i) derives independent economic value, actual or potential, from not being generally known to, and not being readily ascertainable by proper means by, other persons who can obtain economic value from its disclosure or use, and (ii) is the subject of efforts that are reasonable under the circumstances to maintain its secrecy.[35]

The key elements of the trade secret claim are that it protects information that has economic value, is not generally known, and is protected through reasonable efforts to protect its secrecy.[36] First, the subject of the protection must be information, broadly defined. The data or information comprising the bioinformatics databases falls squarely within the broad category of information protected by trade secret law and poses no challenge to protection.

Second, the information must not be generally known to the public. This means that the information in the database in question cannot be available in other, open databases. If the information is not secret, the underlying basis of the trade secret protection disappears.[37] Like copyright, there is no protection for effort expended creating the database, however expensive and time consuming. It is protected only if the database is comprised of information not available elsewhere. If that data is later independently disclosed, then the trade secret protection for the original database is also destroyed. As a result, a contract provision that requires each database user to protect the confidential information in the database will stop reaching any data that has become known elsewhere.

The requirement of secrecy creates a fatal barrier to trade secret protection for many bioinformatics data systems. Only data that is collected in a closed, proprietary system could be subject to trade secret protection. The right to open access for community-based or cross-referenced data systems would undermine any claim of secrecy, which is the critical underlying element for any trade secret case. It is possible that there can be some sharing of databases, provided that each participant agrees to the trade secret obligations, but the law remains unclear on how broad such an agreement could become before the secret would be out and the protection lost.[38]

Third, to retain trade secret status, the value of the data must come, at least in part, from its secrecy. This is typically a low barrier to meet, requiring evidence that the data can be licensed, sold, or used to create a commercial product.[39]

Fourth, the information "is the subject of efforts that are reasonable under the circumstances to maintain its secrecy."[40] Although the steps appropriate are contextual to the situation, minimal reasonable steps for bioinformatics databases will typically include written agreements with each researcher or other person having access to the data that it will be treated as a confidential trade secret; that it will be physically stored in a manner that protects it from unauthorized access or casual observation by third parties; that computer data is protected through encryption; that service level agreements are used that specify obligations for every computer vendor to provide confidentiality regarding the data; and that similar steps are taken to assure that the data will not be seen without authorization. While each of these steps may not be required in every instance, the greater the care, the more likely the court will be able to determine that the information was protected by trade secret law—and the more likely the participants will respect the confidential nature of the data to begin with.

Particularly for bioinformatics data, agreements between researchers and computer service or data storage providers must be explicit that the information being collected is protected as a confidential trade secret and treated accordingly. If the information is stored on a common cloud storage platform, the service level agreement (or terms of use agreement) must specify that the provider understands it is storing confidential data and is taking commercially reasonable steps to protect that data. That service provider must physically and contractually limit access to those servers and preclude its employees from having unauthorized access to the data. Unless the vendor is hosting the software to manipulate the data, it is much better to encrypt the data before it is stored and thus greatly reduce the risk of compromise for the data.

Trade secret protection of nonshared, proprietary data will be available if the data is valuable and the steps described above are followed to protect the data. At the same time, however, this is not consistent with much of the important data sharing dominant in the bioinformatics industry, and therefore it is somewhat limited as a practical solution.

U.S. PATENT PROTECTION FOR ORIGINAL DATA

Processing Methods and Other Inventions

Patent law is designed to protect new, useful, and nonobvious inventions relating to a "process, machine, manufacture, or composition of matter."[41] Human-made genetically created bacteria and other human-made genes and microorganisms are within the subject matter of patents.[42] The U.S. Patent & Trademark Office (USPTO) has adopted regulations that provide guidance and ensure the ongoing patentability of such animal-derived living compositions.[43]

At the same time, "laws of nature, physical phenomena, and abstract ideas"[44] are outside the protection of patent law, as are mental processes.[45] "[I]t is unlikely that one could patent a biological sequence since it may be characterized as a natural phenomenon."[46]

The data in a bioinformatics database, therefore, is not within the scope of patent protection because, as with the copyright analysis, it is not human-made.[47] Moreover, to the extent the database is a compilation of biological sequences, the aggregation of unpatentable data does not change the nature of the data to make it patentable.

There is some ambiguity regarding the patentability of the data schema itself. The basic rule had earlier been that the data schema or structure was not eligible for patent protection.[48] But this determination is apparently open to some reconsideration to the extent that the data structures are embodied in a "computer-readable medium encoded with a data structure [that] defines structural and functional interrelationships between the data structure and the computer software and hardware components which permit the data structure's functionality to be realized."[49] It is also possible that the database interoperability may fall within the subject matter of patent protection.[50]

The ultimate scope of patent protection in this area continues to be in flux. In *Bilski v. Kappos*,[51] the Supreme Court addressed the ongoing tension over the ability to patent business method inventions as well as software. The Court rejected a rule established by the Federal Circuit that an invention is a "process" only if: "(1) it is tied to a particular machine or apparatus, or (2) it transforms a particular article into a different state or thing."[52] The Court rejected this test as being the exclusive test for patentability of a process, but it did not provide significant guidance regarding the proper formulation or limitation of such tests. As a result, these types of business methods and software patents will continue to be within the subject matter of patent law, but it also remains unclear precisely how the Federal Circuit will analyze such applications. This lack of clarity will further diminish the role of patent law for patents related to the database software and structure.

Patent protection is more readily available for inventions derived from the data.[53] For example, "[u]nder appropriate circumstances, it should be possible to obtain commercially significant patent protection for certain aspects of a protein structure."[54] As a result, some data sets with a database "may be protected indirectly by associating the data set with a particular physical structure or use."[55]

As a result, while the data in a bioinformatics database will not be the subject of patent eligibility, there are sufficient possibilities regarding the usage of the data or the structure of the database and its software, that patent eligibility should be taken into account when looking at the database design and contractual provisions. Care should be taken to avoid making publicly available through use, sale, or offer of sale any patent-eligible invention prior to the filing of the patent claim, as these activities will destroy the statutory novelty of the claim. Similarly, if the inventor of the database hopes to seek protection for the database structure, the database cannot be made publicly available except as part of the database development, and then only if subject to restrictions on use so that the exploitation of the database is part of the experimental development of the structure.

In contrast, if the goal is to patent the resulting protein or molecule that may be developed through experimentation in reliance on the data, then the data is not protected by patent law and the patentee need not concern herself with limitations on the data that provided the information critical to inventing the patent. As a result, while patents are critical to bioinformatics financing, they generally do not dictate particular controls over the data themselves.

E.U. DATABASE PROTECTION

In contrast to the highly contract-focused protection of databases in the United States, the European Union has taken a different approach. Generally speaking, in Europe, there is far less ability to rely on terms and conditions promulgated in clickwrap agreements. At the same time, there is a strong recognition regarding the economic importance of databases, so the European Union has developed the sui generis right to protect database works.[56] The directive provides that investment in databases will be protected for 15 years to prevent wholesale use or extraction of the data.[57]

> Member States shall provide for a right for the maker of a database which shows that there has been qualitatively and/or quantitatively a substantial investment in either the obtaining, verification or presentation of the contents to prevent extraction and/or re-utilization of the whole or of a substantial part, evaluated qualitatively and/or quantitatively, of the contents of that database.[58]

At the same time, the owner of the database cannot contractually limit the rights of the lawful users of the database more restrictively.[59] The right is available only to EU member state persons and businesses or those that have reciprocity with the European Union, so U.S. companies hoping to utilize this protection need to avail themselves of European laws through subsidiaries or other means.

The database directive has proven highly controversial and cumbersome. Because the substantial investment prong is vague, there is consternation regarding the point at which the right attaches to data sets. Similarly because data is constantly added to some databases, the 15 years may extend for generations until the database is no longer in use.

Other concerns focus on the impact of the protections provided. The directive protects against substantial extraction and reutilization. Regarding extraction, this may be a quantitatively substantial portion of the database, in reference to the database as a whole.[60] Qualitatively, this "refers to the scale of investment in obtaining, verifying or presenting the particular data being extracted or re-utilized, notwithstanding the question whether that actually represents a quantitatively substantial part of the contents of a database."[61] The nature of information on a bioinformatics database likely leaves only a quantitative protection, since none of the material is intrinsically more valuable than any other material in the database.

Despite the many concerns regarding the sui generis right for data protection, European law is much more squarely protective of the bioinformatics databases than its U.S. counterpart. Large-scale projects will invariably include the necessary time and expense to qualify for protection, and the state laws enforcing the directive will provide significant, proprietary protection to the owners of these databases from large-scale misappropriation.

As with any form of legal protection, it remains within the purview of database owners to forego these protections to foster better community-based information dissemination.[62]

FINAL CONSIDERATIONS REGARDING DATABASE OWNERSHIP AND AUTHORSHIP

The analysis of the various forms of legal protection for bioinformatics databases tends to assume that there is an author or owner of the database. Each of these doctrines is further complicated by the collaborative nature of the database creation. The data itself is typically gathered through research teams, including academic faculty, researchers, students, and employees. Each category may have its own contractual or employment rules regarding ownership of the work produced. In other instances, the teams are from industry. The funding source for the work will often include federal or private grants, with each grant having its own obligations regarding the contractual rules for exploitation of the data. The data sets may be further subject to collaborative agreement between these various teams and their institutions.

The surprise is not that there are myriad data hierarchies and challenges to interactive data mapping; the surprise is that there is any collaboration at all.

For commercial purposes, some research teams will necessarily choose to be wholly proprietary. For others, however, the model that will likely prove most efficient will include a very open development of the data set structure and common language (using a common set of technical standards) enforced through contractual provisions that engage the participants at the institutional level to abide by these standards. Additionally, some commercial software and data will conform to existing international vocabularies or ontologies.

The institutional contractual provisions will include the terms and conditions incorporated into the clickwrap agreement, but the agreement will also include provisions that obligate each participating institution to incorporate by reference the terms of this participation agreement into the terms of its employment and faculty policies. As such, the participation agreement will assure that the agreed-upon level of information sharing, disclosure, and academic freedom is standardized among all participants using the common data.

By contracting at the institutional level, the projects so organized will have the broadest array of researchers and participants all working under the same access and use policies, to maximize the usefulness of the database.

SUMMARY

Although many aspects of intellectual property protection play some role in the protection of bioinformatics databases, U.S. law favors the private ordering of contract protection over the European model of standardized affirmative law governing database information. The data itself is not protected by copyright or patent, though it can be the subject of trade secret law if the standards are met and the data owners are diligent. Even here, the trade secret law is dependent on the enforcement of a comprehensive contracting system that requires all users of the data to agree to the terms of the protection as a condition of their access. It is the contract, even more than the trade secret law, that gives such a system practical value to the owner or owners of the database.

Through the use of a clickwrap agreement on any computer access and written contracts on any noncomputer access, the terms and conditions appropriate to the particular research project can be fully described and effectively enforced.

NOTES

1. M. Scott McBride, *Bioinformatics and Intellectual Property Protection*, 17 Berkeley Tech. L.J. 1331, 1334 (2002).
2. Robert J. Robbins, *Bioinformatics: Essential Infrastructure for Global Biology*, 3 J. Computational Biology 465, 465 n.2 (1996).
3. Robert Stevens, Chris Wroe, Phillip Lord & Carole Goble, *Ontologies in Bioinformatics*, *in* Handbook on Ontology 635 (2004).
4. Robbins, *supra note* 2, at 465.
5. *Id.* at 468.
6. NIH Roadmap, National Centers for Biomedical Computing, Biositemaps, http://www.ncbcs.org/ncbc-biositemaps.html (last visited Oct. 13, 2011).

7. *Id. See also* BIOSITEMAPS, http://biositemaps.org/ (last visited Oct. 26, 2011).

8. 17 U.S.C. § 102(b) (1990) ("In no case does copyright protection for an original work of authorship extend to any idea, procedure, process, system, method of operation, concept, principle, or discovery, regardless of the form in which it is described, explained, illustrated, or embodied in such work."); Feist Publ'ns, Inc. v. Rural Tel. Serv. Co., 499 U.S. 340, 357 (1991).

9. *E.g.*, UNIF. TRADE SECRETS ACT § 1(4) (1985).

10. *See, e.g.*, Vasudevan Software, Inc. v. IBM, 2011 U.S. Dist. LEXIS 7323 (N.D. Cal. Jan. 20, 2011); Timeline, Inc. v. Proclarity Corp., 2006 U.S. Dist. LEXIS 44478 (W.D. Wash. June 29, 2006).

11. Council Directive 96/9/EC, Legal Protection of Databases, art. 9, 1996 O.J. (L 77/20) [hereinafter Database Directive].

12. *See* Andrea M. Matwyshyn, *Resilience: Building Better Users and Fair Trade Practices in Information*, 63 FED. COMM. L.J. 391, 402–03 (2011) ("contract law is currently a more potent framework for legal ordering than is intellectual property law").

13. Ronald Mann & Travis Siebeneicher, *Just One Click: The Reality of Internet Retail Contracting*, 108 COLUM. L. REV. 984, 990 (2008).

14. *Id. See, e.g.*, Register.com, Inc. v. Verio, Inc., 356 F.3d 393 (2d Cir. 2004). *See also* Juliet M. Moringiello & William L. Reynolds, *Survey of the Law of Cyberspace: Electronic Contracting Cases 2005–2006*, 62 BUS. LAW. 195, 201–03 (2006) (describing legal distinction between clickwrap and browsewrap).

15. Genelle I. Belmas & Brian N. Larson, *Clicking Away Your Speech Rights: The Enforceability of Gagwrap Licenses*, 12 COMM. L. & POL'Y 37 (2007); Bragg v. Linden Research, Inc., 487 F. Supp. 2d 593 (E.D. Pa. 2007) (finding arbitration clause and choice-of-forum clauses unconscionable and unenforceable).

16. *Bragg*, 487 F. Supp. 2d at 593.

17. ProCD, Inc. v. Zeidenberg, 86 F.3d 1447 (7th Cir. 1996).

18. Katherine A. Franco, *Protecting Free and Open Source Software: Solutions in the Digital Millennium Copyright Act*, 12 COLUM. SCI. & TECH. L. REV. 160 (2011); Greg R. Vetter, *Open Source and Proprietary Models of Innovation: Beyond Ideology Part III: Open Source and Proprietary Software Development: Slouching Toward Open Innovation: Free and Open Source Software for Electronic Health Information*, 30 WASH. U. J.L. & POL'Y 179 (2009).

19. Health Insurance Portability and Accountability Act of 1996 (HIPAA), Pub. L. No. 104-191, 110 Stat. 1936 (codified at 29 U.S.C. §§ 1181–82 (Supp. III 1997)); Health Information Technology for Economic and Clinical Health Act of 2009, Pub. L. No. 111-5, 123 Stat. 226 (codified at 42 U.S.C. § 17932); *see also* U.S. Dep't of Health & Human Servs., Standards for Electronic Transaction, 65 Fed. Reg. 50,312, 50,312–72 (codified at 45 C.F.R. pts. 160 & 162 (privacy regulations adopted by the Secretary of Health and Human Services).

20. 17 U.S.C. § 102(a) (1990).

21. *Id.* § 102(b).

22. Feist Publ'ns, Inc. v. Rural Tel. Serv. Co., 499 U.S. 340 (1991).

23. *Id.* at 346 (citing *The Trade-Mark Cases*, 100 U.S. 82 (1879)).

24. *Id.* (quoting M. NIMMER & D. NIMMER, NIMMER ON COPYRIGHT § 2.11[A] (1990)).

25. *Id.*

26. Stevens et al., *supra* note 3, at 4.

27. N.Y. Mercantile Exch., Inc. v. Intercontinental Exchange, Inc., 497 F.3d 109, 116–17 (2d Cir. 2007).

28. Kregos v. Associated Press, 937 F.2d 700, 705 (2d Cir. 1991).

29. *Feist*, 499 U.S. at 350.

30. *Id.* at 350–51.

31. 17 U.S.C. § 101 (2010).

32. *See, e.g.*, Utopia Provider Sys. v. Pro-Med Clinical Sys., L.L.C., 596 F.3d 1313 (11th Cir. 2010); *Kregos*, 937 F.2d at 700; Whelan Assocs., Inc. v. Jaslow Dental Lab., Inc., 797 F.2d 1222 (3d Cir. 1986).

33. *See* Eric E. Johnson, *Trade Secret Subject Matter*, 33 HAMLINE L. REV. 545, 550 (2010) (44 states, the District of Columbia, and the U.S. Virgin Islands have adopted the USTA; the other states are covered by common law).

34. 18 U.S.C. § 1832 (2011).

35. UNIF. TRADE SECRETS ACT § 1(4) (1985).

36. *See* Sharon K. Sandeen, *A Contract by Any Other Name Is Still a Contract: Examining the Effectiveness of Trade Secret Clauses to Protect Databases*, 45 IDEA 119, 136–37 (2005).

37. *See id.* at 130.

38. *See, e.g.*, Rockwell Graphic Sys., Inc. v. DEV Indus., Inc., 925 F.2d 174, 179 (7th Cir. 1991) (between engineers, vendors, and customers, thousands of drawings protected by trade secrets were available).

39. *See, e.g.*, Editions Play Bac, S.A. v. Western Publ'g Co., 31 U.S.P.Q. 2d (BNA) 1338, 1342–43 (S.D.N.Y. 1993).

40. UNIF. TRADE SECRETS ACT § 1(4) (1985).

41. 35 U.S.C. § 101 (2011).

42. Diamond v. Chakrabarty, 447 U.S. 303 (1980); Ass'n for Molecular Pathology v. U.S. Patent & Trademark Office, 653 F.3d 1329 (Fed. Cir. 2011).

43. *See* 66 Fed. Reg. 1,092–99 (Jan. 5, 2001).

44. *Chakrabarty*, 447 U.S. at 309; *see* Diamond v. Diehr, 450 U.S. 175, 191–93 (1981).

45. Gottschalk v. Benson, 409 U.S. 63, 67 (1972).

46. McBride, *supra* note 1, at 1342.

47. *Id.*

48. *In re Warmerdam*, 33 F.3d 1354, 1361–62 (Fed. Cir. 1994); U.S. Dep't of Commerce, Manual of Patent Examining Procedure (MPEP) 2106.01 (8th ed., 8th rev. July 2010).

49. MPEP at 2106.01.

50. *Symposium on Bioinformatics and Intellectual Property Law: April 27, 2001, Boston, Massachusetts Patenting and Financing Bioinformatics Inventions*, 8 B.U. J. Sci. & Tech. L. 157, 164–65 (2002) (comments of Thomas C. Meyers).

51. 561 U.S. __, 130 S. Ct. 3218 (2011).

52. *Id.* at 3225 (quoting *In re Bilski*, 545 F.3d 943, 954 (2009)).

53. USPTO, Guidelines for Examination of Applications for Compliance with the Utility Requirement, 66 Fed. Reg. 1092, 1098 (Jan. 5, 2001) (requiring "specific and substantial utility that is credible," to uphold claims).

54. Thomas C. Meyers et al., *Patent Protection for Protein Structures and Databases*, Nature Structural Biology, Nov. 1, 2000, at 950–51, *available at* http://smb.slac.stanford.edu/~ash/nsb_genomics/009_meyer_patent.pdf.

55. *Id.* at 951.

56. Database Directive, *supra* note 11, art. 9; *see* Miriam Bitton, *Exploring European Union Copyright Policy through the Lens of the Database Directive*, 23 Berkeley Tech. L.J. 1411, 1444-46 (2008).

57. Database Directive, *supra* note 11, arts. 7(1), 10(1).

58. *Id.* art. 9.

59. *Id.* art. 16.

60. Mahesh Madhavan, *Copyright versus Database Right of Protection in the UK: The Bioinformatics Bone of Contention*, 9 J. World Intell. Prop. 61, 76 (2006).

61. *Id.*

62. *See* Jasper A. Bovenberg, *Blood, Sweat and Grants "Honest Jim" and the European Database-Right*, 1 Genomics, Soc'y & Pol'y, Aug. 2005, at 1 (raising concerns about the lack of access to data if protected by the database right).

CHAPTER 12

Data Release and Access

Jorge L. Contreras*

INTRODUCTION

Bioinformatics as a discipline is heavily dependent on the availability of vast public collections of data that can be downloaded and analyzed by researchers worldwide. In general, this data has been assembled by publicly and privately funded research projects, beginning with the Human Genome Project (HGP). The availability of this tremendous public data resource is due, in large part, to the data-release policies developed toward the beginning of the HGP, which have been carried forward, in modified form, to the present. These policies impose requirements on both the generators of data (typically the sequencing centers and other laboratories conducting genetic experiments) and the users of that data (i.e., anyone who downloads and/or uses it). This chapter briefly outlines the history of such data-release policies, including those developed by the U.S. National Institutes of Health (NIH), the private sector, and non-U.S. bodies, and provides an overview of the obligations imposed on both data generators and data users.

THE EXPANDING DATA LANDSCAPE

The principal databases for the deposit of genomic sequence data are GenBank, which is administered by the National Center for Biotechnology Information (NCBI)—a division of the NIH's National Library of Medicine; the European Molecular Biology Library (EMBL) in Hinxton, England; and the DNA Data Bank of Japan (DDBJ). NCBI also maintains the RefSeq database, which consolidates and annotates much of the sequence data found in GenBank. In addition to DNA sequence data, genomic studies generate data relating to the association between particular genetic markers and disease risk and other physiological traits. This type of data, which is more complex to record, search, and correlate than the raw sequence data deposited in GenBank, is housed in databases such as the Database of Genotypes and Phenotypes (dbGaP), operated by NIH's National Library of Medicine. dbGaP can also accommodate phenotypic data, which

*This chapter is adapted from Jorge L. Contreras, *Bermuda's Legacy: Policy, Patents, and the Design of the Genome Commons*, 12 MINN. J.L. SCI. & TECH. 61 (2011).

includes elements such as de-identified subject age, ethnicity, weight, demographics, drug exposure, disease state, and behavioral factors, as well as study documentation and statistical results, including linkage and association analyses. Given the potential sensitivity of phenotypic data, dbGaP allows access to data on two levels: open and controlled. Open data access is available to the general public via the Internet and includes nonsensitive summary data, generally in aggregated form. Data from the controlled portion of the database may be accessed only under conditions specified by the data supplier, often requiring certification of the user's identity and research purpose.

The sheer size of the public aggregation of genomic data (which I have termed the "genome commons")[1] is matched only by the breathtaking rate at which it is expanding. Over its decade-long existence, the HGP mapped the 3.2 billion base pairs comprising the human genome. To do so, it sequenced tens of billions of DNA bases (gigabases), creating what was then an unprecedented accumulation of genomic data. By way of comparison, the current 1000 Genomes Project is projected to generate 200,000 gigabases of data—approximately 20,000 times the quantity generated by the HGP.[2] In 2010, the Cancer Genome Atlas project (TCGA) generated data at a rate of 7,300 gigabases per month. Together with the Human Microbiome Project and the 1000 Genomes Project, the total data generation rate from major NIH-funded genome projects in 2010 was nearly 10,000 gigabases per month.[3] And the rate at which DNA sequencing can be accomplished is rapidly increasing. According to one 2011 report, "a single DNA sequencer can now generate in a day what it took 10 years to collect for the Human Genome Project."[4] Statistics like these have led to talk of a "data tsunami" in genomic science, in which the capacity to analyze the vast quantities of data being generated will severely lag behind the rate at which such data is produced.

BERMUDA AND THE ORIGINS OF RAPID GENOMIC DATA RELEASE

In 1992, shortly after the HGP was launched, NIH and the Department of Energy (DOE) developed formal guidelines for the sharing of HGP data.[5] The guidelines required that data generated by the HGP be deposited in GenBank, making it available to all scientists worldwide. Recognizing that the sequencing centers working on the HGP would require time to analyze and prepare publications relating to this data, the policy gave researchers a six-month period before the public release of data was required. The year 1996 marked a turning point for the HGP. Not only was it the year in which sequencing of the human genome was scheduled to begin, it also signaled a sea change in the data-release landscape. That February, approximately 50 project leaders met in Hamilton, Bermuda, to deliberate over the speed with which HGP data should be released to the public, and whether the six-month "holding period" approved in 1992 should continue.[6] The resulting "Bermuda Principles" established that all DNA sequence information from large-scale human genomic sequencing projects should be "freely available and in the public domain in order to encourage research and development and to maximize its benefit to society."[7] They went on to define the method by which such data should be shared, requiring that sequence assemblies greater than one kilobase (kb) in length be released *within 24 hours* after assembly, and that finished annotated sequences should be submitted *immediately* to a public database.

The Bermuda Principles were revolutionary in that they established, for the first time, that data from public genomic projects should be released to the public almost immediately after their generation. Prior to 1996, the position of the National Human Genome Research Institute (NHGRI) with respect to data release and intellectual property was not very different than that of other federal agencies. By and large, such projects required data release after the publication of study results, which often occurred one to two years following the completion of research.[8] But in the negotiations at and leading up to the Bermuda meeting, the scientific community's acknowledgment of the collective norms of data sharing and the public domain, bolstered by the gravitas of several Nobel laureates and other leading figures, seems to have captured the agency's imagination. These norms have since become ingrained as part of NHGRI's basic position treating genomic data as a public good that should be widely available and unencumbered.

DATA RELEASE POST-HGP

The initial draft of the human genome sequence was published by the HGP and a competing private effort in 2001. In 2003, the Wellcome Trust convened a meeting in Fort Lauderdale, Florida, to revisit rapid data-release issues in the "post-genome" world.[9] While the Fort Lauderdale participants "enthusiastically reaffirmed" the 1996 Bermuda Principles, they also expressed concern over the inability of data-generating scientists to study their results and publish analyses prior to the public release of data. The most significant outcome of the Fort Lauderdale meeting was a consensus that the Bermuda Principles should apply to each "community resource project" (CRP), meaning "a research project specifically devised and implemented to create a set of data, reagents or other material whose primary utility will be as a resource for the broad scientific community." Under this definition, the 24-hour rapid release rules of Bermuda would be applicable to large-scale projects generating nonhuman sequence data, other basic genomic data maps, and other collections of complex biological data such as protein structures and gene expression information. In order to effectuate this data-release requirement, funding agencies were urged to designate appropriate efforts as CRPs and to require, as a condition of funding, that rapid, prepublication data release be required in such projects.

Notwithstanding this show of support, the Fort Lauderdale participants acknowledged that rapid, prepublication data release might not be feasible or desirable in all situations, particularly for projects other than CRPs. In particular, the notion of a CRP, the primary goal of which is to generate a particular data set for general scientific use, is often distinguished from "hypothesis-driven" research, in which the investigators' primary goal is to solve a particular scientific question, such as the function of a specific gene or the cause of a specific disease or condition.[10] In hypothesis-driven research, success is often measured by the degree to which a scientific question is answered rather than the completion of a quantifiable data set. Thus, the early release of data generated by such projects would generally be resisted by the data-generating scientists who carefully selected their experiments to test as-yet-unpublished theories. Giving such data away before their theories are published could potentially enable a competing group to "scoop" the originating group, a persistent fear among highly competitive scientists.

In the years following the Fort Lauderdale summit, numerous large-scale genomic research projects were launched with increasingly sophisticated requirements regarding data release. These policies implement their requirements through contractual mechanisms that are more tailored and comprehensive than the broad policy statements of the HGP era. Moreover, increasingly sophisticated database technologies have enabled the provision of differentiated levels of data access, the screening of user applications for data access, and improved tracking of data access and users.

DATA-RELEASE REQUIREMENTS TODAY

Today, all major publicly funded genomic research projects contain detailed data-reporting and data-release requirements. Below is a summary of just a few current programs and their associated requirements.

The Cancer Genome Atlas

In 2006, the National Cancer Institute (NCI) and NHGRI launched a project to catalog genomic changes relating to cancer.[11] The TCGA project, which continues today, generates genomic sequence and related data, but also keeps track of a large amount of clinical data, including patient diagnosis, treatment history, and ongoing status.[12] Due to the specialized nature of the project data, deposits are made both in dbGaP as well as a TCGA-specific database administered by NCI.[13] Given the potential for identifying individual patients from their genomic and phenotypic data, great attention was paid to controlling access to TCGA data. TCGA data is available in an open-access tier and a controlled-access tier.[14] Open-access is provided for data that cannot be aggregated to generate an individually identifiable data set, whereas controlled-access enables researchers to access clinical and individually unique data. Access to the controlled-access data tier requires the user's acknowledgement of a Data Access Certification containing restrictions on research use, security, transferability, and other matters.

The NIH Genome-Wide Association Studies Policy

In response to the growing number of genome-wide association (GWA) studies being conducted and the large amount of genomic data generated by such studies, in August 2007 the NIH released a new policy regarding the generation, protection, and sharing of data generated by all federally funded GWA studies.[15] The NIH Genome-Wide Association Studies (GWAS) policy requires that researchers submit descriptive information about each GWA study for inclusion in the "open access" portion of dbGaP. Researchers are also "strongly encouraged" to submit study results, including phenotypic, exposure, and genotypic data, for inclusion in the "controlled access" portion of the database "as soon as quality control procedures have been completed."[16]

Among the principal concerns raised regarding GWA study data were those surrounding the public release of phenotypic or clinical information that could eventually be traced back to individual subjects.[17] To address this concern, the NIH GWAS policy requires that GWAS data be de-identified in accordance with HIPAA guidelines.[18] Moreover, the data in the controlled-access portion of the database may be released only after approval of the proposed research use by a Data Access Committee,[19] and then only under a signed Data Use Certification. The Data Use Certification requires researchers and their institutions to agree, among other things, to use data only for the approved research purpose, protect data confidentiality, implement appropriate data security measures, not attempt to identify individual data subjects, not sell any data, not share data with third parties, and to report violations to the committee. Finally, the NIH sets forth its position that a request under the federal Freedom of Information Act (FOIA)[20] for the release of individually identifiable GWAS information would constitute an "invasion of personal privacy" under FOIA, and will be denied by NIH.[21]

The NIH GWAS policy was amended in August 2008 following the publication of a scientific paper demonstrating that inferences regarding individual identity could be drawn by analyzing allele frequency data in aggregated genomic data sets and other statistical techniques.[22] Due to concerns relating to potential identification of GWAS subjects, NIH withdrew certain GWAS-generated single nucleotide polymorphism (SNP) data from the publicly accessible portions of dbGaP and certain NCI databases and placed them in the controlled-access portions of these databases.[23]

The NIH GWAS policy addresses the publication priority concerns of data generators by stating an expectation that users of GWAS data refrain from submitting their analyses and conclusions for publication, or otherwise presenting them publicly, during an "exclusivity" period of up to 12 months from the date that the data set is made available. NIH also expresses a "hope" and expectation that "genotype-phenotype associations identified through NIH-supported and NIH-maintained GWAS datasets and their obvious implications will remain available to all investigators, unencumbered by intellectual property claims."[24] It goes on to explain that "[t]he filing of patent applications and/or the enforcement of resultant patents in a manner that might restrict use of NIH-supported genotype-phenotype data could diminish the potential public benefit they could provide."[25] However, in an effort to show some support for patent seekers, the GWAS policy also "encourages patenting of technology suitable for subsequent private investment that may lead to the development of products that address public needs."[26]

ENCODE and modENCODE

In 2007 NIH launched the ENCODE and modENCODE projects to identify functional genomic elements in humans and two simpler model organisms.[27] The ENCODE data-release policy[28] designates the project as a "community resource project" but also recommends a nine-month embargo period during which users of released data are asked not to publish or present results based on that data. The ENCODE Policy distinguishes between published and unpublished data, as well as verified and unverified data, and offers several examples of the data-use implications for different types of studies. The length and complexity of the policy evidences the agency's and the participants' desire for clear guidelines and the avoidance of misunderstandings regarding the release of data, as the diversity of participants, organisms, and data types has expanded dramatically beyond those originally considered by the framers of the Bermuda Principles. This year, ENCODE scientists released a massive series of findings in 30 simultaneously published scientific papers.[29]

The Human Microbiome Project

The Human Microbiome Project (HMP) is a large-scale community resource project initiated in 2008 that is designed to identify and sequence the genomes of a handful of the host of microorganisms inhabiting the human body.[30] While much HMP data is subject to rapid Bermuda-like disclosure requirements, investigators are permitted to withhold certain other data from the public for a period of several months.[31] This hold-back period is intended to permit HMP researchers to analyze and prepare publications on their data before it is released to competing researchers. The reasons that researchers, who are driven by intense competitive pressure to publish and claim credit for discoveries, have pushed for such hold-back periods are clear. However, it also appears, at least in the case of HMP, that NIH has not vigorously advanced the patent deterrent arguments that previously motivated policy decisions during the HGP and its immediate aftermath. Whether the experience of the HMP indicates a new direction for NIH, or simply a minor deviation from its overall policy mission, is not clear.

PRIVATE SECTOR INITIATIVES

In addition to the public-sector projects described above, a number of private-sector research projects have resulted in the creation of large aggregations of genomic data. During the HGP era, privately funded projects such as the Merck Gene Index[32] and the SNP Consortium[33] attracted attention and kudos for releasing significant amounts of data into the public domain. Today, successor projects such as the International Serious Adverse Events (SAE) Consortium (SAEC) continue to fund academic research in targeted areas. SAEC's particular focus is the discovery of genetic markers potentially associated with severe adverse drug reactions.[34] The SAEC works with academic collaborators to collect DNA samples and associated phenotypic data, and then funds GWAS, targeted sequencing, and statistical analyses to identify potential markers and associations of interest. SAEC studies have investigated potential DNA markers relating to drug-induced liver injury, serious skin rash, excessive weight gain, osteonecrosis of the jaw, and other conditions, and all data is released publicly on the SAEC website. The SAEC seeks to minimize patent encumbrances on genetic markers and associations that it identifies via a "protective" patent strategy in which patent applications claiming various DNA markers have been filed to act as prior art against third-party patent filings. Like the public sector policies discussed above, the SAEC imposes various security, research purpose, and nonpatenting restrictions on data that is publicly released.[35] It also secures for data-generating scientists a period of exclusivity (up to 12 months), during which they have sole access to the data.[36] During this time they have the ability to analyze data and prepare papers for publication without the threat of being preempted by competing groups.

Another recent development is the Harvard-led Personal Genome Project (PGP).[37] The PGP, launched in 2008 to significant press coverage, solicits volunteers to submit tissue samples and accompanying phenotypic data. Researchers are then authorized to analyze the submitted samples and publish any resulting genomic information on the PGP website. All such data is released without restriction under the "CC0" Creative Commons copyright waiver.[38] The PGP approach differs markedly from that of the government and privately funded projects described above, in that it dispenses entirely with any attempt to restrict the use of its genomic data. PGP requires its contributors to waive all privacy-related rights when contributing their tissue samples to the project, and gives no preference to use of the data by researchers of any kind.

POLICIES OUTSIDE THE UNITED STATES

Although the HGP and subsequent genome sequencing projects relied on international cooperation and collaboration, the data-release policies adopted by groups outside the United States have differed in material ways from corresponding policies adopted by NIH and NHGRI. In particular, non-U.S. funding agencies have generally exhibited less concern with patenting issues, and have remained more flexible with respect to the time frames both for release of data by data generators and embargo periods on publication for data users. A few examples of recent non-U.S. data-release policies are described next.

Genome Canada

Genome Canada, a participant in the HGP, adopted its first formal data-release policy in 2005.[39] While acknowledging the Fort Lauderdale principles, the Canadian policy does not adopt the 24-hour release requirement set forth in the earlier Bermuda Principles. With respect to data generators, Genome Canada "expects data to be released and shared no later than the original publication date" of the researchers' results, provided that all data must be released "without restriction" by the end of a project. As for patents, Genome Canada "recognizes the need to protect patentable and other proprietary data," and thus requires that data generators' obligation to release data occurs upon the earlier of publication or the filing of a patent application.[40]

Wellcome Trust Case Control Consortium (WTCCC)

The Wellcome Trust is the largest charity in the United Kingdom and the second-largest biomedical research funding charity in the world. Since the beginning of the HGP, the Wellcome Trust has supported genomics initiatives, both through direct funding and through its Sanger Institute in Cambridge, England, a leading sequencing center. In 2006, the Trust funded a large-scale GWA study of seven complex human diseases that was conducted by more than 50 research groups from institutions across the United Kingdom. The study generated a large quantity of data, including aggregated and individual-level genotypic and phenotypic information. Most of this data was made available to the public in accordance with the Fort Lauderdale Principles, and the project designated itself as a CRP.[41] In order to ensure appropriate use of released data, the consortium requires all prospective data users to apply to the consortium's Data Access Committee and sign a written Data Access Agreement.[42] Access to data is granted only to qualified investigators for "appropriate use," as determined by the committee.[43] The data access agreement requires security, acknowledgment, transfer, and use restrictions comparable to those found in the Genetic Association Information Network (GAIN)[44] and other recent policies.[45] It also includes some restrictions that are specific to the study samples, such as a prohibition on any use of data from the 1958 British Birth Cohort for commercial purposes. The agreement does not, however, contain any specific embargo on publication or any restriction on patenting activity.

UK Medical Research Council

In 2008, the UK's Medical Research Council (MRC) released a comprehensive set of guidelines surrounding release of data from MRC-funded research.[46] In a set of broad "data access principles," the MRC announced that data generated by publicly funded research would be a public good and, as such, "must be made available for new research purposes in a timely, responsible manner." Following the reasoning behind the Fort Lauderdale principles, the MRC states that access to data "must balance the interests of data creators, custodians, users and data subjects," and acknowledges that "limited, defined" periods of exclusive use "will often be justifiable." Beyond these broad pronouncements, however, the MRC gives little specific guidance with respect to the timing or manner of data release. Like the WTCCC guidelines, the MRC guidelines place a high value on formal, written agreements to govern the relationships between data generators and data users. Such agreements "must" be used if restrictions on the use of data are to be imposed, and are "particularly important" when publication rights and intellectual property are implicated.[47] By and large, however, the MRC allows individual parties to define the specific requirements of their data-sharing agreements and does not attempt to impose overarching rules regarding the timing of data release.

BEYOND GENOMICS

The success and broad adoption of genomics data-release policies incorporating the Bermuda and Fort Lauderdale principles have led scientists in related fields to consider the adoption of analogous principles in their own research. One prominent example occurred in 2008, when the NCI convened a meeting of proteomics researchers in Amsterdam to "identify and address potential roadblocks to rapid and open access to [proteomics] data."[48] Participants identified technical, infrastructure, and policy challenges to the rapid release of proteomic data. Technical challenges included the wide variety of disparate platforms and techniques used to generate proteomic data, making "raw" data from experimental instruments difficult to interpret by scientists unfamiliar with, or lacking access to, the instruments used to generate the data. Proteomics also lacks the established public database infrastructure of genomics. Whereas DNA sequence data can be deposited readily in GenBank, the EMBL, or DDBJ, and is often deposited in all three, there is no common public data repository for proteomic data, and existing proteomic databases suffer from inconsistent and sometimes incompatible data formats. Finally, unlike genomics, in which the entire field focused for several years on the single HGP project, proteomics research lacks a unifying policy core and proteomics-focused journals have each developed their own, sometimes inconsistent, guidelines for data submission.

Notwithstanding these difficulties, the Amsterdam participants articulated six data-release and data-sharing principles that reflect the spirit of the Bermuda and Fort Lauderdale principles: (1) timing (should depend on the nature of the effort generating the data, but should in no event be later than publication or, for community resource projects, following appropriate quality-assurance procedures), (2) comprehensiveness (full raw data sets should be released together with associated metadata and quality data), (3) format (standardized formats are encouraged), (4) deposition to repositories (central repositories for proteomic data should be established), (5) quality metrics (central repositories should develop metrics for assessing data quality), and (6) responsibility (scientists, funding agencies, and journals share responsibility for ensuring adherence to community data-release standards).

In 2009, more than one hundred scientists, journal editors, legal scholars, and representatives of governmental and private funding agencies met in Toronto to assess the current state of rapid prepublication data release and the applicability of the Bermuda Principles in projects well beyond the generation of genomic sequence data. The participants reaffirmed a general community commitment to rapid prepublication data release, expanding the scope of projects as to which these principles should apply to all biomedical datasets that "[have] broad utility, are large in scale . . . and are 'reference' in character." Specifically, they cited, in addition to genomic and proteomic studies, structural chemistry, metabolomics, and RNAi datasets, as well as annotated clinical resources such as cohorts, tissue banks, and case-control studies.

The expansion of rapid prepublication data-release principles beyond genomics and proteomics projects, which often have as their ultimate goal the generation of a large data set, to these other areas necessarily raises issues concerning the appropriateness of rapid data release in hypothesis-driven research. Accordingly, the Toronto participants concurred that, while funding agencies should *require* rapid prepublication data release for "broad utility" projects, rapid data release "should not be mandated" for projects that are generally hypothesis-driven. The Toronto participants also addressed the priority concerns of data generators versus data users, observing anecdotally that in many cases data users have, in fact, published papers based on publicly released data sets *before* the publication of the data generators' papers analyzing the data sets themselves, and that this situation caused no "serious damage" to the data generators' subsequent publications. Nevertheless, the participants acknowledged the acceptability of a "protected period" during which data users could be restricted from publishing on released data sets, cautioning, however, that this period should never exceed one year. The Toronto participants produced a set of "best practices" embodying these principles and applying them to the three constituencies originally identified in Fort Lauderdale—funding agencies, data generators, and data users—as well as to the scientific journals, which were urged to monitor and provide guidance relating to data-release issues.

NOTES

1. Jorge L. Contreras, *Bermuda's Legacy: Policy, Patents, and the Design of the Genome Commons*, 12 Minn. J.L. Sci. & Tech. 63 (2011).
2. National Human Genome Research Institute, *1000 Genomes Project data available on Amazon Cloud*, NIH News (Mar. 29, 2012), http://www.nih.gov/news/health/mar2012/nhgri-29.htm.
3. Brad Ozenberger, *TCGA: A Future Arrived*, The Cancer Genome Atlas, http://cancergenome.nih.gov/researchhighlights/leadershipupdate/ozenberger (last visited Sept. 10, 2012).
4. Elizabeth Pennisi, *Will Computers Crash Genomics?*, 331 Sci. 666, 666 (2011).
5. NIH, *DOE Guidelines Encourage Sharing of Data, Resources*, Human Genome News (Oak Ridge Nat'l Laboratory, Oak Ridge, Tenn.), Jan. 1993, at 4.
6. *International Large-Scale Sequencing Meeting*, Human Genome News (Oak Ridge Nat'l Laboratory, Oak Ridge, Tenn.), Apr.–June 1996, at 19.
7. *Summary of Principles Agreed at the First International Strategy Meeting on Human Genome Sequencing*, U.S. Department of Energy Genome Program, http://www.ornl.gov/sci/techresources/Human_Genome/research/bermuda.shtml (last visited Oct. 28, 2010).
8. National Academy of Sciences, Ensuring the Integrity, Accessibility, and Stewardship of Research Data in the Digital Age 64 (2009); National Research Council, Sharing Publication-Related Data and Materials: Responsibilities of Authorship in the Life Sciences 75 (2003); National Research Council, Bits of Power—Issues in Global Access to Scientific Data 80–82 (1997); J.H. Reichman & Paul F. Uhlir, *A Contractually Reconstructed Research Commons for Scientific Data in a Highly Protectionist Intellectual Property Environment*, 66 Law & Contemp. Probs. 315, 335 (2003).
9. Report of Meeting organized by the Wellcome Trust, *Sharing Data from Large-Scale Biological Research Projects: A System of Tripartite Responsibility* (Jan. 14–15, 2003), *available at* http://www.genome.gov/Pages/Research/WellcomeReport0303.pdf.
10. Jane Kaye et al., *Data Sharing in Genomics—Re-shaping Scientific Practice*, 10 Nature Rev. Genetics 331, 332 box 1 (2009).
11. Francis S. Collins & Anna D. Barker, *Mapping the Cancer Genome*, Sci. Am., Mar. 2007, at 50.
12. *Types of Data*, The Cancer Genome Atlas Data Portal, http://cancergenome.nih.gov/dataportal/data/about/types/clinical/ (last visited Oct. 28, 2010).
13. *Data Use Certification* 1, The Cancer Genome Atlas Pilot Project (Feb. 22, 2010), http://dbgap.ncbi.nlm.nih.gov/aa/wga.cgi?page=DUC&view_pdf&stacc=phs000178.v1.p1.
14. *Data Access*, The Cancer Genome Atlas Data Portal, http://cancergenome.nih.gov/dataportal/data/access/ (last visited Oct. 28, 2010).
15. Policy for Sharing of Data Obtained in NIH Supported or Conducted Genome-Wide Association Studies (GWAS), 72 Fed. Reg. 49,290, 49,294–97 (Aug. 28, 2007) [hereinafter NIH GWAS Policy]; *Modifications to Genome-Wide Association Studies (GWAS) Data Access*, Nat'l Inst. of Health (Aug. 28, 2008), http://grants.nih.gov/grants/gwas/data_sharing_policy_modifications_20080828.pdf [hereinafter *Modifications to GWAS Data Access*].
16. NIH GWAS Policy, *supra* note 15, at 49,295.
17. *Id.* at 49,292.
18. *Id.* at 49,295.
19. *Id.* at 49,296.
20. 5 U.S.C. § 552 (2006).
21. FOIA Exemption 6, 5 U.S.C. § 552(b)(6).
22. Modifications to GWAS Data Access, *supra* note 15; Nils Homer et al., *Resolving Individuals Contributing Trace Amounts of DNA to Highly Complex Mixtures Using High-Density SNP Genotyping Microarrays*, PLoS Genetics (Aug. 2008), http://www.plosgenetics.org/article/info%3Adoi%2F10.1371%2Fjournal.pgen.1000167.
23. *Modification to GWAS Data Access, supra* note 15.
24. NIH GWAS Policy, *supra* note 15, at 49,296.
25. *Id.* at 49,297.

26. *Id.* at 49,296.
27. Susan E. Celniker et al., *Unlocking the Secrets of the Genome*, 459 NATURE 927 (2009).
28. ENCODE Consortia, DATA RELEASE, DATA USE, AND PUBLICATION POLICIES (2008), *available at* http://www.genome.gov/Pages/Research/ENCODE/ENCODEDataReleasePolicyFinal2008.pdf.
29. B. Maher, *The Human Encyclopaedia*, 489 NATURE 46 (2012).
30. Peter J. Turnbaugh et al., *The Human Microbiome Project*, 449 NATURE 804 (2007); *Human Microbiome Project Awards Funds for Technology Development, Data Analysis and Ethical Research*, NIH NEWS (Oct. 7, 2008), http://www.genome.gov/27528386.
31. *HMP Data Release and Resource Sharing Guidelines for Human Microbiome Project Data Production Grants*, NIH COMMON FUND, http://commonfund.nih.gov/hmp/datareleaseguidelines.asp.
32. Press Release, Merck & Co., Inc., First Installment of Merck Gene Index Data Released to Public Databases: Cooperative Effort Promises to Speed Scientific Understanding of the Human Genome (Feb. 10, 1995), *available at* http://www.bio.net/bionet/mm/bionews/1995-February/001794.html.
33. Arthur Holden, *The SNP Consortium: Summary of a Private Consortium Effort to Develop an Applied Map of the Human Genome*, 32 BIOTECHNIQUES 22 (2002).
34. *iSAEC's Background and Organizational Structure*, INT'L SAE CONSORTIUM http://www.saeconsortium.org/ (last visited Oct. 28, 2010).
35. *See* Jorge L. Contreras, Aris Floratos & Arthur L. Holden, *The International Serious Adverse Events Consortium's Data Sharing Model*, 31 NATURE BIOTECH. 17–19 (2013).
36. Int'l SAE Consortium Ltd., DATA RELEASE AND INTELLECTUAL PROPERTY POLICY (last amended Nov. 5, 2009) (on file with author).
37. Personal Genome Project, www.personalgenomes.org/mission.html (last visited Sept. 10, 2012); *see generally* MISHA ANGRIST, HERE IS A HUMAN BEING: AT THE DAWN OF PERSONAL GENOMICS (2010) (describing the author's participation in the Personal Genome Project, together with its history and personalities).
38. CREATIVE COMMONS, http://creativecommons.org/publicdomain/zero/1.0/legalcode (last visited Sept. 10, 2012).
39. Genome Canada, DATA RELEASE AND RESOURCE SHARING (Sept. 18, 2008), *available at* http://www.genomecanada.ca/medias/PDF/EN/DataReleaseandResourceSharingPolicy.pdf.
40. *Id.*
41. *Publications Policy*, WELLCOME TRUST CASE CONTROL CONSORTIUM, https://www.wtccc.org.uk/ccc1/publications_policy_ext.shtml (last visited Oct. 26, 2010).
42. *WTCCC: Access to Genotype Data*, WELLCOME TRUST CASE CONTROL CONSORTIUM, https://www.wtccc.org.uk/docs/CDAC_Guidelines_and_Information_July09.pdf (last visited Oct. 26, 2010); *Data Access Agreement*, THE WELLCOME TRUST CASE CONTROL CONSORTIUM, https://www.wtccc.org.uk/docs/Data_Access_Agreement_v15.pdf (last visited Oct. 26, 2010).
43. *WTCCC: Access to Genotype Data*, *supra* note 42, § 4.
44. *See generally* The GAIN Collaborative Research Group, *New models of collaboration in genome-wide association studies: The Genetic Association Information Network*, 39 NATURE GENETICS 1045 (2007) (explaining the selection and characteristics of initial GAIN studies, the structure of GAIN, and defining who has access to GAIN data).
45. *Data Access Agreement*, *supra* note 42.
46. *Principles for Access to, and Use of, MRC Funded Research Data*, MEDICAL RESEARCH COUNCIL, http://www.mrc.ac.uk/consumption/groups/public/documents/content/mrc003759.pdf.
47. *Id.* at 5 para. 8.
48. Henry Rodriguez et al., *Recommendations From the 2008 International Summit on Proteomics Data Release and Sharing Policy: A Summit Report*, 8 J. PROTEOMICS RES. 3689 (2009).

CHAPTER 13

Information Security and Privacy Challenges in Bioinformatics

Lucy L. Thomson and Arthur E. Peabody Jr.

A TORRENT OF DATA

In bioinformatics—where biology, computer science, and information technology merge—major developments in the field of molecular biology, coupled with advances in genomic technologies, have led to an explosive growth in the biological information generated by the scientific community. The growth in the collection of genomic information has, in turn, led to a requirement for computerized databases to store, organize, and index the data and for specialized tools to view and analyze it. The vast amount of genomic information and the use of technology to analyze the data represent both challenges and risks—namely, the use of genomic data to discover new treatments for disease and disabilities while preventing unauthorized use of the data that violates the privacy of patients.

From a privacy and security perspective, two questions are critical. What sensitive and personally identifiable data are included in the bioinformatics databases and in the databases that serve as the sources of data for those databases? How must those data be secured?[1] When data such as patient name and contact information; health care records containing medical diagnoses and other personal health information; insurance coverage and reimbursement data; and financial or employment information are found in a database, then applicable federal, state, and sometimes international laws require that personally identifiable information be protected with appropriate information security.

In light of government initiatives to create electronic health records, data breaches and the resulting invasions of privacy illustrate starkly the challenges ahead for medical professionals, researchers, database owners and custodians, and policymakers. They underscore the need for renewed vigilance to address the problem of data breaches among health care organizations at a time when genomic information in large databases is growing exponentially.

In the data-rich environment of patient health information, privacy and information security are among the most critical issues that must be addressed in the growing field of bioinformatics. As information systems become both more extensive and more complex, the risks to personal information grow.[2] Under a variety of domestic and international laws, researchers, health care providers, and business owners have legal obligations to protect the privacy and security of the individually identifiable health information they collect, process, store, and share.

The risks to individuals of identity theft, fraud, and violations of privacy are becoming more serious as large amounts of information of greater sensitivity become accessible in electronic form to people and organizations that may not have an appropriate and adequate level of security protection. Adequate information system security is necessary to protect privacy.

Medical identity theft is particularly devastating for individuals whose medical information has been stolen.[3] Maintaining patient trust and confidence is essential to health care organizations since the impact of data breaches can be so severe. Individuals whose identities have been stolen face an uphill battle to address the myriad of problems that follow, including closing accounts, canceling fraudulent debts, changing identification numbers, removing incorrect information, and restoring credit. Medical identity theft is much more serious because once personal medical information is disclosed, there is no going back—there are no accounts to change or numbers to reissue. Discrimination, embarrassment and humiliation, and permanent damage from disclosure of the details of an individual's medical condition will follow.[4]

IN THIS CHAPTER

Citing the trend to collect, maintain, share, and transmit increasing amounts of personally identifiable patient health information by electronic means, this chapter reviews the legal requirements for privacy and information security in the Health Insurance Portability and Accountability Act (HIPAA), the Health Information Technology for Economic and Clinical Health (HITECH) Act, and the Genetic Information Nondiscrimination Act (GINA). Identifying a number of significant breaches of protected health information and the sources of threats to privacy, the chapter focuses on preventing data breaches. It outlines the methodology for conducting a risk assessment—the essential building block to protect personal health records—and continuous monitoring, a strategy to continuously assess the security of information systems, and raises concerns as to whether present steps to protect health information are fully adequate to secure data and afford patient privacy.[5]

BIOLOGICAL DATABASES

Research into genomic interpretation is now progressing on a broad front and the collection of genetic and other medical data is growing by leaps and bounds.[6] Genomics information can have a significant impact on the prevention, diagnosis, and treatment of disease.[7] Recent studies have demonstrated the usefulness of genomics for diagnosing disease and guiding treatment in the clinic. Genetic diagnostic testing has resulted in patients receiving improved and tailored treatments based on their unique genetic structure.[8]

Genetic tests are already used to assess the risk of breast[9] and ovarian cancers,[10] diagnose recessive diseases such as cystic fibrosis, determine drug dosages based on individual patient metabolism, and identify therapeutic options for treating lung and breast tumors, melanoma, and leukemia. Whole-genome sequencing may make it possible for tissue typing information to be incorporated into networks for organ sharing, which provides better and more rapid identification of donor-recipient matches.[11]

Creating a system that could efficiently diagnose patients with rare diseases offers a way of implementing the tools and procedures that will later be used in more widespread applications of genomic medicine. Sequences that reveal rare metabolic disorders would produce many other findings of interest, which clinicians could then use in routine care.[12]

Individual genetic information is increasingly stored in public, private, and governmental databases, both in the United States and throughout the world.[13] These databases are typically used for clinical purposes or research.[14] Indeed, commentators have observed that "exponentially growing biological and bio-informatics data sets" pose a number of challenges to the medical profession, researchers, and society as a whole.[15]

Various organizations have sought to create lists or catalog the number, variety, and kind of bioinformatics databases in existence. The main publicly available, genome database sites include the National Center for Biotechnology Information (NCBI),[16] the European Bioinformatics Institute,[17] and the DNA Data Bank of Japan.[18]

GenBank is the National Institutes of Health (NIH) database maintained and distributed by NCBI that stores all known public DNA sequences. Sequence data are submitted to GenBank by individual scientists from around the world, as well as from the large centers involved in the Human Genome Project.[19] The number of DNA sequences stored in the GenBank database, from all organisms, has increased recently and continues to grow at a rapid rate.[20] GenBank is an international collaborative project, with partners located at the European Bioinformatics Institute in the United Kingdom and the National Institute of Genetics in Japan.[21]

The European Molecular Biology Laboratory (EMBL) Nucleotide Sequence Database (also known as EMBL-Bank) constitutes Europe's primary nucleotide sequence resource and is maintained by the European Bioinformatics Institute. The main sources for DNA and RNA sequences are direct submissions from individual researchers, genome sequencing projects, and patent applications. The DDBJ or the DNA Data Bank of Japan, the sole nucleotide sequence databank in Asia, is officially certified to collect nucleotide sequences from researchers.

Each of the three groups—the NCBI, European Bioinformatics Institute, and the DDBJ—collects a portion of the total sequence data reported worldwide, and all new and updated database entries are exchanged between the groups on a daily basis.[22] As such, the three databanks possess and share virtually the same data at any given time. The virtually unified database is called the International Nucleotide Sequence Database (INSD). DDBJ collects sequence data mainly from Japanese researchers, but it accepts data from researchers in other countries.

Many other projects and databases address or collect information about intra-species variation, "open-source" and prediction, 3D nucleic structure, structural classification of RNA, proteins, expression atlases, and model organisms,[23] and all genomes of humans or other species sequenced or otherwise analyzed to diagnose genetic maladies, predict clinical correlations, or conduct medical research. These entities are both public and private. Numerous universities in the United States and across the world offer degrees in bioinformatics and conduct active research programs that utilize bioinformatics.[24] Bioinformatics will be essential in building a genomic basis for clinical work because of the huge amounts of data involved. Because bioinformatics is becoming a recognized specialty area, the American Board of Medical Specialties will recognize clinical informatics as a subspecialty in 2013.

Private companies are also deeply immersed in bioinformatics.[25] For example, Genome, an international corporation, is developing "innovative bioinformatics products" and "custom research solutions for corporate, government, and academic laboratories."[26] In this circumstance, many private organizations possess vast amounts of genetic information.[27]

Due to a variety of factors, including the contribution of data by many entities, ownership of the raw data is "rarely a clear-cut issue."[28] When copyright disputes have arisen in this context, most data have been found to be in the "public domain."[29] The NIH National Center for Biotechnology Information accepts only de-identified information into its GenBank database.[30] While the databases maintained by the European Bioinformatics Institute and the DNA Data Bank of Japan contain no direct statements that they accept only de-identified information, their close collaboration and daily sharing of information suggest that they follow the same policy.[31] However, in order to conduct the initial research or gather the genetic information about individuals—at least in the United States—consent, in general, is required and is referenced on the GenBank website.

Advocates believe that the "synergy" obtained from the joint efforts of the genomics and the clinical and health care disciplines will enhance the personal health and well-being of U.S. citizens.[32] Skeptics point to potential discrimination and threats to individual privacy.[33] In addition, inappropriate or unauthorized access to genetic information may lead to personal stigma, psychological stress, and tensions among family members.[34]

THE FUTURE OF BIOINFORMATICS (MEDICAL INFORMATICS) AND CLINICAL PRACTICE

Medical practitioners and technology experts from across the world see a growing trend where well-informed health care decisions are enhanced by the use of the best available medical research evidence. Trends on the horizon include these possibilities:[35]

- A universal health care information technology system will include both genetic and clinical information.
- Barriers to data sharing will be reduced. Research into genomic interpretation is now progressing on a broad front, but clinical and family information will be needed to understand the effects of genetic variants. Furthermore, this information is needed in the public domain so that people can use it.
- Medical records systems will feature greater interoperability so that information relevant to care follows people throughout their lives and genomic information can act as an incubator for innovation.
- Health care providers will be able to check a patient's status on a special website, such as a social media site, where such information is posted routinely.
- "Crowdsourcing" means that patients will share their own health-related information. For example, the Association of Cancer Online Resources has about 18,000 cancer patients who share information on diagnoses and therapeutic interventions.[36]
- An effective method of reinterpreting and updating genomic information throughout a patient's life will allow physicians to gather clinical information and write an order for updated interpretation of the genome based on the validated information that has accrued since the patient's last encounter with the health care system.[37]
- The role of clinical trials will be changed by genomics medicine—clinical trial designs must be adaptive and engage patients.

Horizon 2020 is the European Union's framework program for research and innovation.[38] A primary goal of the initiative is to ensure that Europe has world-class research infrastructures, including e-infrastructures, accessible to all researchers in Europe and beyond.[39]

Medical professionals and researchers see only a broader horizon for bioinformatics. The role of both genetics and computers will expand to enhance the quality of medical care across the world.

LEGAL REQUIREMENTS AND IMPLICATIONS

The HIPAA Privacy Rule

The HIPAA Privacy Rule (Privacy Rule)[40] establishes a category of health information, defined as protected health information (PHI)[41] that a covered entity[42] may use or disclose to others only in certain circumstances and under various conditions. PHI is a subset of "personally identifiable information."[43] The Privacy Rule applies to individually identifiable health information created or maintained by a covered entity.[44] Medical research data is regulated by a number of federal statutes, including HIPAA. Physicians provide PHI for research purposes either directly or through hospitals and their laboratories or independent laboratories. Hospitals and laboratories, including university-based hospitals and laboratories, collect health care data for research. Any of these entities may send health care data to governmental databases maintained to facilitate a broad range of medical research activities.

Researchers are not themselves covered entities unless they are also health care providers and engage in any of the covered transactions. For example, physicians who conduct clinical studies or administer experimental therapies to patients during the course of a study must comply with the Privacy Rule if they are employed or affiliated with a covered entity, e.g., hospitals, affiliated medical schools, and laboratories. Indeed, researchers and their employers may be bound by HIPAA if they receive their information from a covered entity. A covered entity's use or disclosure of PHI to create a research database or repository and the use or disclosure of PHI from the database or repository for a future research purpose are each considered a separate research activity under the Privacy Rule—each requiring authorization or some waiver or invocation of an exception pursuant to HIPAA. On the other hand, researchers using de-identified health care information where the individually identifiable information has been removed need not comply with any of the HIPAA requirements.

It is important for researchers and research entities to be aware of the HIPAA requirements and how the Privacy Rule affects them in the various types of organizations in which they operate and what requirements the Privacy Rule may impose on their research activities.[45] A research organization must ask itself at least four questions. Is it a covered entity,[46] e.g., is it a health care plan, health care clearinghouse, or a health care provider that transmits health information electronically? Is its relationship with a covered entity such that it meets the HIPAA requirements for a "business associate"?[47] Does it assist or perform a function regulated by HIPAA on behalf of a covered entity, including the use or disclosure of individually identifiable health information? Finally, does the information it receives meet the definition of PHI, or has it been de-identified within the meaning of the applicable regulations (i.e., personally identifiable information has been removed)?[48] Under the HIPAA Privacy and Security Rules, the custodian of the data may have legal responsibility either as a covered entity or as a business associate for maintaining the security and protecting the privacy of individuals whose data are contained in the databases. Liability may also be shared by the "owner" of the data or the entity that submitted the material to the research database on the same basis.

At the same time, it must be recognized that hospitals, clinics, and other health care providers can gather information about their patients for treatment, payment, and health care operations, and place this information in their own databases without authorization from the patient. These entities also routinely report PHI to government agencies for "disease surveillance, disease prevention, and other public health purposes."[49] Various state health-mandated reporting requirements also fall outside of the HIPAA privacy requirements.[50]

In addition to the Privacy Rule, other regulations may apply as well. For instance, individual records held by covered entities that are also alcohol and substance-abuse treatment providers are protected by the Confidentiality of Alcohol and Substance Abuse Patient Records regulation.[51] Also, the HHS and U.S. Food and Drug Administration (FDA) Protection of Human Subjects regulations may apply to health services research.[52] In addition, if health-related research involves electronic PHI (e-PHI), covered entities must also consider the requirements of the HIPAA Security Rule.[53]

Use and Disclosure of PHI for Research

The Privacy Rule permits covered entities to use or disclose PHI for research purposes, either with an individual's specific written permission, termed an "authorization," or without it, if certain conditions are met.[54] A covered entity is permitted to use or disclose PHI for research purposes if it:

- Obtains the individual's authorization for the research use or disclosure of PHI as specified under 45 C.F.R. § 164.508 of the Privacy Rule.[55]
- Obtains satisfactory documentation of an Institutional Review Board (IRB) or Privacy Board's waiver of the authorization requirement that satisfies 45 C.F.R. § 164.512(i) of the Privacy Rule.
- Obtains satisfactory documentation of an IRB or Privacy Board's alteration of the authorization requirement as well as the altered authorization from the individual.
- Uses or discloses PHI for reviews preparatory to research with representations from the researcher that satisfy 45 C.F.R. § 164.512(i)(1)(ii) of the Privacy Rule.
- Uses or discloses PHI for research solely on decedents' PHI with representations from the researcher that satisfy 45 C.F.R. § 164.512(i)(1)(iii) of the Privacy Rule.
- Provides a limited data set and enters into a data use agreement with the recipient as specified under 45 C.F.R. § 164.514(e) of the Privacy Rule.
- Uses or discloses information that is de-identified in accordance with the standards set by the Privacy Rule at 45 C.F.R. § 164.514(a)–(c) (in which case the health information is no longer PHI).

When authorization is not practical, the Privacy Rule provides a number of alternatives. The most common is the use of "de-identified data sets." To avoid HIPAA requirements, PHI may be de-identified in two ways:

(1) In accordance with 45 C.F.R. § 164.514(b)(2), the data may be "stripped" of the 18 identifiers set forth in the regulations, e.g., name, identification numbers, etc.[56] This method is commonly used and will serve to de-identify the data unless the covered entity has actual knowledge that it would be possible to use the remaining information to identify the individual.

(2) A qualified statistician, using generally accepted statistical and scientific principles, may determine that there is only a very small risk that the information might be used alone or in combination with other information to identify the individual. The methods utilized by the statistician must be documented.

Moreover, in some cases, de-identified data may lack critical information needed for health services research (e.g., nine-digit ZIP codes or dates of treatment). When such indirect identifiers are needed for the research, a covered entity may provide the data to a researcher as a limited data set. No authorization or waiver or alteration of authorization by an IRB or Privacy Board is required for a covered entity to use or disclose a limited data set.

Limited data sets are data sets stripped of certain direct identifiers that are specified in the Privacy Rule.[57] Limited data sets may be used or disclosed only for public health, research, or health care operations purposes. Because limited data sets contain certain identifiers, they are *not* de-identified information under the Privacy Rule. Importantly, unlike de-identified data, PHI in limited data sets may include the following: addresses other than street name or street address or post office boxes, all elements of dates (such as admission and discharge dates), and unique codes or identifiers not listed as direct identifiers in 45 C.F.R. § 164.514(e).

Before disclosing a limited data set to a researcher, a covered entity must enter into a data use agreement with the researcher. Among other requirements set forth in 45 C.F.R. § 164.514(e)(4) of the Privacy Rule, the data use agreement must identify who will receive the limited data set, establish how the data may be used and disclosed by the recipient, and provide assurances that the data will be protected. If a covered entity learns that the researcher has violated this agreement, the entity must take reasonable steps to end or repair the violation. If such steps are unsuccessful, the entity must stop disclosing PHI to the researcher and report the problem to the Department of Health and Human Services (HHS) Office for Civil Rights.

In addition, for some types of research, de-identified information or a limited data set may not be sufficient for the research purposes. It also may be impracticable for researchers to obtain written authorization from research participants, for example, for some research conducted on existing databases or repositories where no contact information is available. To address these situations, the Privacy Rule contains criteria for waiving or altering the authorization requirement by an IRB or another review body, called a Privacy Board. The Privacy Rule permits a covered entity to use or disclose PHI for research purposes without authorization if the covered entity receives proper documentation that an IRB or Privacy Board has granted a waiver (or an alteration) of the authorization requirement for the research use or disclosure of PHI. The Privacy Rule establishes criteria to be used by an IRB or Privacy Board in approving a waiver or alteration of the authorization requirement.[58]

There are other provisions permitting access to PHI for research involving a decedent,[59] reviews preparatory to research,[60] and grandfathering provisions. For all methods of disclosure where authorization is not the basis for the use or disclosure of PHI, the use or disclosure is limited to the minimum necessary amount to accomplish the research[61] and the individual has a right to an accounting of all uses or disclosures made.[62]

Finally, questions have been raised as to whether de-identification of data is adequate to protect patient privacy.

Is De-Identification Reliable as a Means to Protect Privacy?

Under the aegis of HIPAA and the Privacy Act of 1974, large databases rely on representations that they accept only de-identified data to protect the privacy of those patients whose data fill their databases. Yet an increasing number of scientists and scholars are raising questions as to whether de-identified data can be re-identified in many cases.

Commentators have observed that despite using various measures to de-identify health records, it is possible to re-identify a significant number of them by using voter registration records, hospital discharge records, and other information in commercially available databases.[63] Another observes that up to 87 percent of Americans can be identified by combining their ZIP code, birth date, and sex.[64] They opine that the HIPAA criteria for de-identification are inadequate to assure anonymity.[65] In a now-famous incident, the Massachusetts Group Insurance Commission released "anonymized" data on state employees showing hospital visits. Then Massachusetts Governor William Weld assured the public that their privacy was fully protected. Shortly thereafter, the hospital records of Governor Weld were identified.[66]

Further questions have been raised as to whether de-identification is adequate to protect the privacy rights of individuals whose genetic information forms the core of the largest bioinformatics databases.[67] Studies indicate that genetic information can be linked using publicly available databases to other genetic samples and that it is possible to identify an individual's genomic attribute data in a pooled or aggregated data sample.[68] Further, "knowledge bases and machine learning methods" can relate de-identified genomic sequences to identified clinical records.[69]

Many genome-based research projects include familial relationships, such as pedigrees, with genomic data records. At least one commentator has observed that when family relations are disclosed, such as in the form of a pedigree, the potential exists for an invasion of privacy. The danger arises because "information about family relations resides in various formats external to the de-identified data."[70]

These critics argue that currently accepted methods for de-identification of data need to be evaluated and, left unchanged, pose serious threats to the privacy of millions of Americans whose genomic data has been entered into massive databases both here and elsewhere throughout the world.[71]

Genetic Information Nondiscrimination Act (GINA)

In recent years, discrimination based on genetic information has become a major concern for patients and physicians. Genetic data may be considered more sensitive than the privacy of clinical data, as it could lead to discrimination not only to the individual but also to relatives who have not been genetically tested. Further, predictions about disease conditions or medical risks can be made for an individual who has been genetically tested. As such, the risk of genetic discrimination is real for many individuals.

Recent court cases have identified the challenges and risks associated with the collection and use of genetic information. Courts have recognized the sensitive nature of genetic information, and recent decisions reflect a perceived need for additional protection of this information.[72] In addition, state legislatures have enacted laws to regulate the use of genetic information.[73] Various work groups, committees, and institutes studied the prevalence of discrimination and made recommendations, all of which culminated in the passage of the Genetic Information Nondiscrimination Act (GINA).[74]

Since 2008, GINA[75] has been the legal standard for the collection, use, and disclosure of genetic information. Considering new developments in the availability of genetic information and its use for treatment, health care professionals must be cognizant of the requirements of this statute when gathering genetic and other health care information. Although

GINA specifically addresses genetic information, health care professionals need to be aware of how this legislation amended current laws, including HIPAA.[76]

In order to prohibit discrimination based on genetic information, Title I of GINA sets forth requirements regarding the use of such information in making decisions about health insurance coverage. Specifically, health plans and group health insurers are prohibited from using genetic information to adjust premium amounts for a group covered under a plan; from requiring or requesting that a family member or individual take a genetic test; and from requiring, requesting, or purchasing genetic information for underwriting purposes or with respect to any individual before he or she enrolls in the plan. In addition, GINA expanded the definition of "protected health information" (PHI) to include genetic information. Consequently, genetic information is now protected by HIPAA.

GINA defines genetic information as "information about (i) such individual's genetic tests, (ii) the genetic tests of family members of such individual, and (iii) the manifestation of a disease or disorder in family members of such individual."[77] In addition, "any request for, or receipt of, genetic services, or participation in clinical research which includes genetic services, by such individual or any family member of such individuals" is defined as genetic information.[78] Furthermore, any genetic information relative to a fetus carried by a pregnant woman or to an embryo legally held by an individual or family member using assisted reproductive technology is also considered protected. "Genetic tests" include an analysis of human DNA, RNA, chromosomes, proteins, or metabolites that detects genotypes, mutations, or chromosomal changes.[79] "Genetic services" include genetic counseling or genetic education.[80]

Health plan providers and insurers will continue to have a legitimate need for the minimum information necessary for payment of health care claims, including claims for genetic tests. The documents required will vary depending on the type of claim; however, it can be assumed that genetic information may be included for payment of claims related to testing, treatment, or prophylactic procedures. In these instances, the provided information will be protected as PHI and subject to HIPAA.

Patients' family medical histories are traditionally collected during annual wellness exams, as well as during enrollment prescreening physicals. Considering the broad definition of "genetic information," collecting family history may present a risk to medical professionals; therefore, appropriate security measures are mandatory. Family history information obtained regarding a patient currently enrolled in a wellness program is also specifically protected and may not be used for underwriting purposes.

Title II of GINA applies to all employers, employment agencies, labor organizations, and training programs. GINA specifically prohibits discrimination against employees based on genetic information in hiring, termination, referral, or other decisions regarding compensation, terms, conditions, or privileges of employment. If an employer possesses any genetic information in an employee file, that information must be kept separate and treated as a confidential medical record.

The GINA/HIPAA Interface

GINA amended HIPAA to include genetic information in the definition of personal health information.[81] In addition, the HIPAA regulations include penalties for the misuse of genetic information. Business associate agreements must be drafted to comply with the guidelines and limitations established by GINA. Health plan policies and procedures, as well as their Notice of Privacy Practices, likewise must reflect the mandates of GINA.

With the increased use of electronic medical records that may contain genetic information, the risk of a data breach—and a violation of HIPAA or HITECH—is heightened. Health care professionals must identify the sources of genetic information in patient records. These sources range from documentation provided in the family history or physical examination to laboratory results. All sources must be protected by adequate security measures and closely monitored throughout the release of information to ensure proper protection.

Under current legal requirements, the de-identification of PHI in the three major databases maintained by the National Center for Biotechnology Information, the European Bioinformatics Institute, and the DNA Data Bank of Japan may effectively protect these entities—and the individuals whose genetic information is reflected in their files—from invasions of privacy, and in the case of NIH and its American partners, from violations of HIPAA.[82]

HIPAA Coverage of "Business Associates"

Although HIPAA was not intended to impede medical research using records within databases and other repositories, the growth of medical research databases presents genuine threats to the privacy of individuals.[83] The Privacy Rule allows covered entities and health plans to disclose protected health information to "business associates." The Omnibus Final Rule issued by HHS on January 25, 2013, makes most provisions of the Privacy Rule directly applicable to business associates. In general, a business associate is not authorized to use or further disclose protected health information in a manner that would violate the requirements of the Privacy Rule if done by the covered entity.

A "business associate" is a person or entity that performs certain functions or activities involving the use or disclosure of protected health information on behalf of, or provides services to, a covered entity.[84] For example, many laboratories are business associates of other health care providers. Where a physician and an entity maintain a bioinformatics database from which information is obtained to assist in the care of a patient, whether present or future, the entity operating the database may be a business associate of the physician. In other words, the continuing relationship between the physician and the bioinformatics database entity to provide ongoing medical care to a patient may accord the database provider business associate status.

The commentary to the Omnibus Final Rule, however, makes clear that a person or entity is a business associate only where it is conducting a function regulated by HIPAA, for example, payment health care operations or services listed in the definition of business associate and has access to PHI.[85] The commentary reasons that an external researcher may not be a business associate merely as a result of its research activities—even if the covered entity has hired the researcher to conduct research.[86] However, those involved in a function or activity covered by the statute would be business associates, for example, those creating a de-identified or limited data set for the covered entity.[87] No researcher working with or under a contract with a covered entity, however, should parse these definitions too finely—the goal of HIPAA and the new Omnibus Final Rule is to protect PHI.

Business associates must safeguard appropriately the protected health information they receive or create on behalf of a covered entity. In most cases, the satisfactory assurances must be in writing, whether in the form of a contract or other agreement between the covered entity and the business associate.

The HIPAA Security Rule

While the major bioinformatics databases are reported to contain only de-identified patient data, the broader health care environment in which health care records are created, used, shared, and transmitted contains numerous data repositories with a wealth of individual electronic patient records and PHI.

The national goal of developing an electronic health record for all Americans by 2014, consisting of a broad array of the most sensitive types of information about individuals, has stimulated the creation of electronic PHI that can be transmitted to other entities and stored in information repositories around the world. These electronic medical records contain detailed information such as patient contact information, Social Security number, date of birth, height, weight, sex, race; medical, psychological, and dental information; laboratory and diagnostic test results; registries; social, economic, and demographic data; health services utilization; insurance and hospital cost data; employer, conditions of work environment, and exposure to hazardous substances/compounds; information pertaining to stored biologic specimens (including blood, urine, tissue, and genetic materials); genetic information; characteristics and activities of health care providers and educators and trainers; and associated correspondence.

Individuals and organizations involved in the health care and research environment who collect sensitive personal and medical information and store, use, and share it electronically are required under HIPAA to protect and safeguard those records.

The HIPAA Security Rule sets national standards for the security of electronic protected health information (e-PHI).[88] This rule should be followed by entities maintaining databases containing PHI.[89] The Security Rule requires covered entities to maintain reasonable and appropriate administrative, technical, and physical safeguards for protecting e-PHI. Specifically, covered entities must:

- Ensure the confidentiality, integrity, and availability of all e-PHI they create, receive, maintain, or transmit,
- Identify and protect against reasonably anticipated threats to the security or integrity of the information,
- Protect against reasonably anticipated, impermissible uses or disclosures, and
- Ensure compliance by their workforce.

The Security Rule is designed to be flexible and scalable so a covered entity can implement policies, procedures, and technologies that are appropriate for the entity's particular size, organizational structure, and risks to consumers' e-PHI. Therefore, when a covered entity is deciding which security measures to use, the Rule does not dictate those measures but requires the covered entity to consider:

- Its size, complexity, and capabilities,
- Its technical, hardware, and software infrastructure,
- The costs of security measures, and
- The likelihood and possible impact of potential risks to e-PHI.

Finally, the Security Rule does not mandate the use of any specific technology, such as encryption. However, HHS guidance issued following the passage of HITECH provides a "safe harbor" for PHI that is either encrypted or has been destroyed. If a data breach occurs involving encrypted PHI, there is no violation of the statute.[90]

The most significant risks to privacy or of a data breach appear to be at the physician's office, where PHI is collected routinely, as well as in organizations such as hospitals and health plans that maintain large databases containing health records of millions of patients.[91] At the physician level, sensitive data are being entered into electronic medical records, and with the growth of health information exchanges, transferred from one medical doctor or health care entity to another. In these circumstances, it is clear that HIPAA applies—PHI is being transmitted electronically. It appears that physicians and hospitals (including affiliated and independent laboratories) are gathering health care information, including PHI.

Databases containing PHI that have not been de-identified or made subject to a limited data set are subject to HIPAA—with a myriad of exceptions designed to facilitate beneficial medical research. The creation of a database or repository for research and the use or disclosure of PHI from a research database or repository, including those maintained by universities and their affiliated medical schools, is a research activity under the Privacy Rule.

Large government databases, federal as well as state and local, may or may not contain PHI. The largest genomics database, e.g., GenBank, maintained by NIH, contains only de-identified data and, as such, poses the most minimal risks to individuals.[92] In the large government databases that accept only de-identified data, risks to privacy may be lessened. A number of state databases containing health care claims data have experienced data breaches that exposed millions of patient records.[93] The application of the Privacy Rule to specific databases must be considered on an individual basis. While NIH may not be covered by HIPAA,[94] the agency is nonetheless bound to follow the Federal Privacy Act of 1974[95] and FISMA requirements. FISMA requires compliance with administrative, technical, and operational security standards for electronic records.[96]

HITECH—The Federal Data Breach Notification Law

Congress enacted the first federal data breach notification law as part of the Health Information Technology for Economic and Clinical Health Act, known as the HITECH Act,[97] a law whose primary purpose is to promote health information technology, including electronic health records. HITECH and HIPAA[98] are designed to protect personal health information from unauthorized access and use. Pursuant to the statutory scheme, covered entities must protect this individually identifiable health information by encrypting such information or utilizing another means to protect the information. And if unauthorized access occurs, they must notify individuals of the breach in a timely manner. With the collection of significant personal information about individuals, both identified and de-identified, all those involved with bioinformatics need to be prepared to address the requirements of the recently enacted federal data breach notification statute.

In enacting HITECH, Congress has endeavored to respond to the need to protect electronic health records and address widespread invasions of privacy caused by breaches of medical and other individually identifiable health information. HITECH requires entities covered by the statute to notify individuals whose PHI has been breached. Generally, PHI includes all medical records, including electronic medical records, and other information reflecting individually identifiable health information maintained by an entity or its business associates. Both HHS and the Federal Trade Commission (FTC) have issued final interim rules or regulations designed to implement HITECH requirements.[99] Professionals in the field of bioinformatics will be confronted with a complex regulatory scheme.

HHS issued a Omnibus Final Rule, 78 Fed. Reg. 5566 (January 25, 2013), defining a data breach as the unauthorized acquisition, access, use, or disclosure of protected health information that compromises the security and privacy of such information. A breach is presumed to have occurred unless, pursuant to a risk assessment, the covered entity or business associate can demonstrate that there is a low probability that the PHI has been compromised.[100] This standard replaces the "harm standard" enunciated in the proposed final rule, that is, that the breach must "pose a significant risk of financial, reputational, or other harm to the individual."[101] In abandoning the harm standard, the agency acknowledges that breach notification is necessary in all cases except where the covered entity or business associate can demonstrate a low probability that the PHI has been compromised. A risk assessment must be conducted to determine whether the PHI has been compromised; the Final Interim Rule required a risk assessment to ascertain whether the individual had been harmed. The focus of the risk assessment has been changed to reflect the revised standard enunciated in the Omnibus Final Rule.

The HHS Omnibus Final Rule provides a "safe harbor" that, if followed, nullifies the notification requirement. Covered entities that encrypt PHI consistent with standards promulgated by National Institute of Standards and Technology (NIST) or destroy the information are not required to comply with the notification requirement in the event of a data breach. The means of notification required vary depending on the extent of the breach, i.e., the number of individuals whose PHI has been breached. In some cases, only individuals must be notified; in "larger" breaches, the media and the secretary of HHS must also be notified.

NIST publications present a broad array of guidance that a covered entity may utilize or follow to protect individually identifiable health information, including encryption. No single "standard" will ensure the security of an information system. While HHS focuses on the need for encryption, no information system can be secure absent an appropriate security design, selection implementation of security controls based on a risk assessment, and timely maintenance and periodic monitoring of the system—all issues professionals in the field of bioinformatics will need to address in order to ensure compliance with this federal mandate.

The FTC has also issued regulations pursuant to HITECH that cover vendors; its jurisdiction is limited to non-HIPAA covered entities. As such, the FTC regulations do not cover electronic health records made and utilized in a physician's office. The primary focus of the regulations is the security of personal health records maintained by individuals on vendor websites.

State Data Breach Notification Laws

Forty-six states, the District of Columbia, Puerto Rico, and the Virgin Islands have enacted legislation requiring notification of individuals whose personal information may have been affected by a security breach.[102] California was the first state to enact a data breach notification law, and other states have followed its lead. Generally, the laws require that any business in possession of sensitive personal information about a covered individual must disclose a breach of that information. Many health care providers, researchers, and other entities that maintain databases with personal health records, including the owners and custodians of large bioinformatics databases, are covered by these state laws. Each state law defines the type of personal information covered by the law, who must be notified, when notice must be provided, and the content and form of the notice.

The definition of a breach varies from state to state. Generally, the statutes require notice to affected individuals following the unauthorized acquisition of unencrypted computerized data that compromises the security, confidentiality, or integrity of personal information about individuals. However, notice is not required in several states unless there is a reasonable basis to believe the breach will result in substantial harm or inconvenience to the affected individual. In most cases, businesses may avoid the notification requirements if the breach involved information that has been encrypted, provided the encryption key has not been compromised.

Some states have responded to the massive data breaches by expanding their laws to provide some type of information security to protect sensitive personal information—not merely to report when data have been hacked or stolen. Nevada and Massachusetts have enacted statutes requiring substantial security measures. Massachusetts requires any firm conducting business with state residents to develop a comprehensive information security program. They must employ encryption, adopt a risk-based approach to ensuring security, and protect against data leakage. Nevada has taken the additional step of requiring holders of credit card and other personal information in computerized information systems to comply with the professional standards promulgated by the Payment Card Industry Council. In addition, businesses are required to encrypt all transmissions of personal information.

In California, data custodians must implement reasonable security procedures and practices. New Jersey law limits disclosure and transmission of Social Security numbers over the Internet. Maryland requires organizations to employ reasonable security procedures. In recent years, many states have expanded the scope of their data breach laws, set additional requirements related to notification, and changed the penalties for those responsible for breaches.

HEALTH CARE DATA BREACHES AND INFORMATION SECURITY: ADDRESSING THREATS AND RISKS TO PATIENT DATA

The Need for Risk Assessment

Ensuring the confidentiality, integrity, availability, and privacy of data in bioinformatics databases is fundamental to protecting the personal information of patients. Information security is not a mystery—it is based on a systematic assessment of threats and risks that are present in a particular information system.

Federal laws governing the privacy and security of patient health records require that government and private-sector organizations conduct a risk assessment as part of their security management processes.[103] Based on a risk assessment, appropriate security controls can be selected, implemented, and continuously monitored, so that risks and vulnerabilities are reduced to a reasonable and appropriate level.[104]

The principal goal of an organization's risk management process should be to protect the *organization and its ability to perform its mission*, not just its IT assets. A risk analysis process[105] includes, but is not limited to, the following activities:

(1) Evaluate the likelihood and impact of potential risks to e-PHI;
(2) Implement appropriate security measures to address the risks identified in the risk analysis;
(3) Document the chosen security measures and, where required, the rationale for adopting those measures; and
(4) Maintain continuous, reasonable, and appropriate security protections.

Risk analysis should be an ongoing process, in which an organization regularly reviews its records to track access to e-PHI and detect security incidents, periodically evaluates the effectiveness of security measures put in place, and regularly reevaluates potential risks to e-PHI.

Health Care Data Breaches on the Rise—Threats and Vulnerabilities Illustrated

Massive data breaches are occurring with alarming frequency. An analysis of data breaches by industry should provide a wakeup call for the health care industry—health care/medical

Table 13.1 Percent of Data Breaches by Industry Sector

Industry	2012 (Jan.–June)
Business	40.0
Health/Medical	28.2
Educational	15.9
Government/Military	11.2
Banking/Credit/Financial	7.0

data breaches now account for more than 25 percent of all data breaches. Health care data breaches have increased steadily from fourth place in 2006–2009 to second place behind only the business sector in 2010 and 2011 (see Table 13.1).

Failed security has resulted in massive data breaches that lead to the loss or compromise of millions of personally identifiable health care records.[106] In order to prevent data breaches, it is essential to analyze and understand the root causes of the security failures and develop a specific plan to address them. Analysis of the types of breaches that have occurred is very illuminating—for the conclusion that most of these breaches did not have to happen. To protect one of the most valuable and vulnerable assets of all organizations—data and information—is not only good business practice, it will avoid the high costs associated with responding to data breaches, potential liability, negative press, embarrassment, and ultimately loss of trust of patients and health care providers.[107]

In almost all cases, data breaches that occurred could have been prevented by proper planning and the correct security design and implementation of appropriate security safeguards. Individuals and organizations alike, particularly health care providers and the owners and custodians of large bioinformatics databases, have cause for serious concern. In today's digital world, threats to information systems are evidenced almost everywhere a computer or laptop, smartphone, USB drive, or other electronic device is operating. The proliferation of mobile devices and wireless technology that enable mobile health (m-health) as well as an expanding array of software applications present many vulnerable points in the flow of sensitive data in computer networks. With the increasing emphasis on genetic research and the transmission of genomic data from various sources to databases and perhaps ultimately to one of the major genome databases, the risk of a breach of genetic data is genuine.

In just one month in mid-2012, three major data breaches resulted in the potential exposure of more than 1.3 million personal health records with a myriad of medical information, including genome data. A closer look reveals the types of lax security representative of health care data breaches generally over the past five years.

Lost and stolen computers and mobile devices—Emory Healthcare, Inc., in Georgia revealed that it lost ten computer disks containing patient Social Security numbers for about 315,000 surgical patients.[108] Research universities are also involved in these types of breaches.[109]

Hacker attacks—The Utah Department of Technology Services (DTS) publicized the breach of a server that houses Medicaid claims of 780,000 patients. It was suspected that hackers operating out of Eastern Europe were able to circumvent the security system of the server where the data were stored.[110]

Malicious insiders—The South Carolina Department of Health and Human Services revealed a major Medicaid data breach affecting the protected health information of more than 228,000 Medicaid beneficiaries. The department discovered that an employee, since terminated, transferred spreadsheets with personal information (including contact information, dates of birth, ID numbers, and Social Security numbers) to a personal e-mail account.[111]

Several months earlier in 2011, Sutter Physicians Services and Sutter Medical Foundation of Sacramento, California, announced one of the largest health care data breaches ever—a password-protected, unencrypted desktop computer containing the health records of 3.3 million patients was stolen from their administrative offices. The computer contained patient contact information, dates of birth, telephone numbers, e-mail addresses, medical record numbers, and health insurance information.[112]

Causes of Data Breaches

Mobile Devices

Data breaches are pervasive because they are not only the result of hacker attacks, but they are also caused by sloppy security practices. More than one third of data breaches resulted from the theft or loss of laptops, computers, hard drives, backup tapes, PDAs, or other portable media containing unencrypted personal information. Some of the largest breaches involving lost and stolen unencrypted mobile devices are identified in Table 13.2 below.

Table 13.2 Health Care Data Breaches Caused by Lost or Stolen Mobile Devices[113]

Date	Company	Records Breached	Lost/Stolen Devices
September 2011	TRICARE and SAIC	4.6 million	Backup tapes stolen
November 2011	Sutter Physicians Service and Foundation, CA	4.2 million	Desktop computer stolen
March 2011	HealthNet and IBM, CA	2 million	Server drives lost
February 2011	Jacobi Medical Center, Bronx, NY	1.7 million	Backup tapes stolen
October 2011	Nemours, DE	1.6 million	Three backup tapes missing
November 2009	Health Net, CT	1.5 million	Portable hard drives lost
October 2009	BlueCross BlueShield, TN	1 million	Hard drives missing
September 2009	Oklahoma Dept. of Human Services	1 million	Computer stolen
October 2009	BlueCross BlueShield Highmark	850,000	Laptop stolen
March 2012	California Dept. of Child Support	800,000	Computer devices lost
April 2011	Eisenhower Medical Center, CA	514,000	Computer stolen
April 2012	Emory University Hospital GA	315,000	Ten backup tapes stolen
March 2011	Cord Blood Registry	300,000	Backup tapes stolen
April 2011	Oklahoma Dept. of Health	133,000	Laptop stolen
April 2011	Mid-State Medical Center, CT	93,500	Hard drives lost
February 2011	St. Francis Hospital, OK	84,000	Computer stolen
November 2011	Neurological Institute of Savannah	63,000	Laptop stolen
March 2012	Howard University Hospital, DC	34,500	Laptop stolen

These breaches illustrate the sheer magnitude of the exposure of personal records and the potential damage to millions of individuals. In light of available security measures and their widespread acceptance within the information security community, there is no excuse for health care organizations that collect and store PHI to fail to fulfill their duty to protect personal patient information.

Data breaches can be prevented by encrypting the data on the devices, enforcing procedures that do not permit individuals to transport sensitive data on moveable media, keeping careful track of the devices, and having the highest standards and requirements for couriers to move backup tapes and CDs to offsite storage facilities or from one location to the another. The HHS Omnibus Final Rule provides a "safe harbor" so covered entities that encrypt PHI consistent with standards promulgated by NIST or destroy the information are not required to comply with the HITECH notification requirement if a data breach occurs.

The Identity Theft Resource Center has analyzed the problem this way:

> This is 100% avoidable, either through use of encryption, or other safety measures. Laptops, portable storage devices and briefcases full of files, outside of the workplace, are still "breaches waiting to happen." With tiered permissions, truncation, redaction and other recording tools, PHI can be left where it belongs—behind encrypted walls at the workplace.[114]

Malicious Insiders

Malicious insider attacks are on the rise and pose a serious threat to the security of sensitive information. These activities accounted for another third of the breaches. Individuals who abused administrative privileges led to a significant number of breaches, and security failures such as not suspending system access for terminated employees caused many breaches.

By mid-2012, a range of 17 to 29 insider breaches was reported by organizations tracking data breaches.[115] In 2011, 56 insider breaches were reported, down from 102 insider breaches in 2010. During 2007, 2008, and 2009, a total of 214 data breaches caused by insiders were reported. About one quarter of the reported insider breaches were perpetrated by employees who had some type of job change—they were fired, resigned, were newly hired, or changed roles within the organization.

Defending an organization's perimeter from external attack does not protect against valuable information seeping out because of insider malfeasance, whether that behavior is characterized as malicious, mischievous, or ignorant/accidental.[116] Carnegie Mellon University has published a list of best practices[117] to minimize the risk of insider attacks. Recent incidents and law enforcement activity against malicious insiders include:

- A federal grand jury indicted a surgical instrument technician at University of Pittsburgh Medical Center (UPMC) Shadyside for selling patients' names, birth dates, and Social Security numbers; he pled guilty.[118]
- In April 2012, a former manager of the Head Injury Association on Long Island, New York, was indicted for stealing the identities of patients. A 48-count indictment alleged that he used the names and Social Security numbers of patients to e-file fraudulent tax returns and obtained over $200,000 in federal and state tax refunds.[119]
- In January 2012, United Healthcare (UHC) discovered employee access to information in a database that UHC learned was unauthorized. The information that may have been accessed regarding patients included name, Social Security number, telephone number, date of birth, address, and Medicare Healthcare Insurance Number.[120]
- In November 2011, several employees of the Berkeley Heart Lab in California were found to have accessed patient information without authorization and to have taken the data to a competitor.[121]

Databases are vulnerable to attacks from both insiders and hackers. These vulnerabilities are well documented and include the use of default and weak passwords, failure to patch known vulnerabilities, misconfigurations, and excessive privileges granted to users. Specific security measures must be taken to protect personal data in biometrics, bioinformatics, and other health care databases.[122]

Inadvertent Postings at Websites and Other Disclosures

Through careless handling, personal data, particularly Social Security numbers, are often exposed in postal mailings. In addition, data may also be posted on websites where it is accessible to individuals who are not intended to have access to it. For example, in February 2012, the St. Joseph Health System alerted more than 31,000 patients in California that their PHI records may have been searchable on the Internet.[123]

Peer-to-peer (P2P) file-sharing technology creates a significant risk of a breach when it is installed on the same computer as personal information; when the P2P file-sharing software is not configured properly, sensitive records not intended for sharing may be accessible to anyone on the P2P network. In several cases, personal health records were breached this way.[124]

Cloud Computing Risks

Because of the increased flexibility and efficiency afforded by computing resources available on demand, cloud environments may be ideally suited to processing genomic data. Large amounts of computing power are needed for research within short time periods. However, cloud computing introduces IT security and privacy risks related to outsourcing the administration and physical control of sensitive data to a third-party vendor, and maintenance of the data on shared computing platforms, risks that will need to be carefully evaluated and addressed by entities that intend to put genomic data of patients in the cloud.[125]

Achieving Optimal Network Security through Continuous Monitoring

To protect against massive data breaches, it is clear that the health care industry, as well as companies in the retail, education, and financial sectors, and government agencies, must take immediate action to strengthen their security posture. Government officials and private-sector executives have reached a consensus that a prioritized approach to information security is required. Continuous monitoring of systems and networks is essential to protect against the broad range of serious threats to information systems.[126]

At the direction of the U.S. Congress, NIST is developing a common information security framework for the federal government and its contractors.[127] Robust continuous monitoring will provide senior executives with the information necessary to make cost-effective, risk-based decisions.[128] While NIST guidance has been developed for government agencies, it is equally appropriate for private-sector organizations. Significantly, NIST has issued guidance tailored for health care organizations.[129]

Information security continuous monitoring is a critical part of organization-wide risk management. To properly support an organization's risk management framework, security must be incorporated into the architecture and design of the organization's information systems and supporting information technology (IT) assets. Continuous monitoring provides an understanding of threats and threat activities, evaluating the security impact of actual and proposed changes to the system; assessing all security controls; collecting, correlating, and analyzing security-related information; providing actionable communication of security status across all levels of the organization; and active management of risk by organizational officials.

Continuous monitoring is designed to provide meaningful, actionable intelligence and reporting—instead of merely collecting data. Cybersecurity situational awareness will inform executives about the threat, vulnerability, and compliance posture of the system, and provide information about incidents that should be investigated. Potential threats must be investigated and targeted attacks can be detected in advance or addressed as they occur, enabling a highly proactive security posture. The result of continuous monitoring will ensure that all the information necessary to address security incidents and potential attacks is readily available for analysis and review. The objective is to address the multitude of security threats and risks in a timely, disciplined, and structured fashion. The tenets of continuous monitoring apply to bioinformatics databases.

Action Plan to Address Health Care Data Breaches

Data breaches can and must be prevented. Organizations that collect, use, store, and share patient information must accept responsibility for protecting PHI and ensuring that it is not compromised by hackers or malicious insiders or inadvertently accessed, lost, or stolen.

The intense focus on response after a data breach has occurred is very costly to organizations and does little to remediate the losses of the individuals whose personal information has exposed them to identity theft or fraud. Business owners, data custodians, and their attorneys in the private sector and in government around the globe, particularly those involved in health care, must redouble their efforts to understand the risks of collecting, storing, and sharing sensitive and personal information; the many ways data can be breached; and the tools available to protect the data. They must develop the expertise to avoid these breaches and to address them appropriately if and when they occur. This includes an appreciation of the fact that no single solution is appropriate for every organization. Information security must be tailored based on the type(s) and sensitivity of data to be protected, the potential risks to the data, and the nature of the information system of the organization, including the hardware, software, applications, and interconnections. As well, business owners and data custodians must understand that implementation and periodic auditing are as important as the initial security controls that are put in place.

Privacy and security must be considered when system design requirements are being developed and decisions are being made about the data and information that will be collected, and how it will be used in the system and shared. Adoption of best practices, including continuous monitoring and adherence to the tenets of the HIPAA Privacy and Security Rules, will ensure that health care organizations can maintain information security and avoid data breaches—and the massive invasions of privacy that result.

Is Protecting the Privacy and Security of Personal Health Records an Attainable Goal?

It is not necessary or appropriate to assume that electronic medical records cannot be protected from data breaches. Health care providers, researchers, and other health care entities have a legal responsibility to protect medical records from unauthorized access, whether malicious or unintentional, by both insiders and hackers. To do this, organizations must have appropriate privacy and security policies in place to protect the data, as well as risk-based management, administrative, and technical controls necessary to ensure information security. Securing personal patient information in large health care databases must be approached systematically. By conducting a risk assessment and prioritizing steps to secure an information system, the majority of privacy and security risks can be addressed.

As the amount of genomic information collected in electronic form continues to grow, the imperative of information security will only grow stronger.

NOTES

1. "Sensitive" refers to information that requires some degree of heightened protection from unauthorized access, use, disclosure, disruption, modification, or destruction because of the nature of the information, e.g., personally identifiable information and proprietary commercial information. Under HIPAA, "individually identifiable health information" is information, including demographic data, that relates to:

> the individual's past, present or future physical or mental health or condition, the provision of health care to the individual, or the past, present, or future payment for the provision of health care to the individual, and that identifies the individual or for which there is a reasonable basis to believe can be used to identify the individual. Individually identifiable health information includes many common identifiers (e.g., name, address, birth date, Social Security number).

45 C.F.R. § 160.103.

2. *See* Presidential Commission for the Study of Bioethical Issues, *PRIVACY and PROGRESS in Whole Genome Sequencing* (Oct. 2012), http://bioethics.gov/cms/node/764 ("this report focuses on recommendations aimed at pursuing and securing the public benefits anticipated from whole genome sequencing while minimizing the potential privacy risks to individuals").

3. World Privacy Forum, *Medical Identity Theft: The Information Crime That Can Kill You* (2006), http://worldprivacyforum.org/pdf/wpf_excum_medidtheft2006.pdf.

4. Enormous costs of breach response and liability to organizations for data breaches are growing—lawsuits have been filed in the case of virtually every major data breach. For example, in the recent breach involving Sutter Health, two lawsuits were filed. One class action lawsuit alleges that Sutter Health was negligent in safeguarding its computers and data and then did not notify the millions of patients whose data were missing within the time required by law. This suit seeks $1,000 for each of the more than three million class member and attorneys' fees.

5. *See* LUCY THOMSON, DATA BREACH AND ENCRYPTION HANDBOOK (2012). Some of the ideas in this chapter were first presented in this book.

6. NIH National Center for Biotechnology Information (NCBI), *A Science Primer*, http://www.ncbi.nlm.nih.gov/About/primer/bioinformatics.html.

7. *See* Christian Baumgartner et al., *Bioinformatics-Driven Search for Metabolic Biomarkers in Disease*, J. CLINICAL BIOINFORMATICS (January 2011), *available at* http://www.jclinbioinformatics.com/content/1/1/2/.

8. *Id.*

9. *See* BREASTCANCER.ORG, http://www.breastcancer.org/symptoms/testing/genetic/.

10. *See* National Cancer Institute, *BRCA1 and BRCA2: Cancer Risk and Genetic Testing*, http://www.cancer.gov/cancertopics/factsheet/Risk/BRCA (last visited Nov. 5, 2012).

11. *See* Thomas Peters, *Kidney Transplant: What It Means*, http://www.aakp.org/aakp-library/kidney-transplant-matching/ (last visited Oct. 28, 2012).

12. *See* Genetic Home Reference, *What Are the Benefits of Genetic Testing?*, http://ghr.nlm.nih.gov/handbook/testing/benefits; *see also* National Academy of Sciences, Symposium on Integrating Large-Scale Genomic Information into Clinical Practice, *available at* http://www.nap.edu/catalog.php?record_id=13256p.

13. Kenneth Goodman, *Toward Striking a Balance in Bioinformatics*, VIRTUAL MENTOR, March 2001.

14. *Id.*

15. Prema Sethi & Kimberly Theodos, *Translational Bioinformatics and Healthcare Informatics: Computational and Ethical Challenges*, *available at* http://prespectives.ahima.org.

16. ENTREZ, The Life Sciences Search Engine, http://www.ncbi.nlm.nih.gov/Entrez.

17. The EBI hosts two databases, one for nucleotide sequences and one for protein sequences, and it is the successor to the European Nucleotide Library. EMBL-EBI, *About EMBL-EBI and Our Role In Europe*, www.ebi.ac.uk/information/About_EBI (last visited Nov. 5, 2012).

18. DNA DATA BANK OF JAPAN, http://www.ddbj.nig.ac.jp (last visited Nov. 5, 2012).

19. Completed in 2003, the Human Genome Project (HGP) was a 13-year project coordinated by the U.S. Department of Energy and NIH. HUMAN GENOME PROJECT INFORMATION, *available at* http://www.ornl.gov/sci/techresources/Human_Genome/home.shtml.

20. *See* NCBI, *From Sequence to Survival: The Search to Understand Disease*, http://www.ncbi.nlm.nih.gov/About/glance/story_discovery.html (last visited Nov. 9, 2012).

21. *Id.*

22. *See* DNA Data Bank of Japan, *supra* note 18.

23. *Id.*

24. Cincinnati Children's Pediatric Informatics, *Degree Programs in Bioinformatics and Computational Biology*, http://anil.cchmc.org/University.html (last visited Nov. 9, 2012).

25. See http://dir.yahoo.com/Business_and_Economy/Business_to_Business/Scientific/Biology/Biotechnology/Genetics/?o=a for a listing of American corporations engaged in genetic testing and research.

26. Genome International, http://www.genome.com (last visited Nov. 9, 2012).

27. *See* GenProMarkers™, Inc., http://genpromarkers.com/ (last visited Nov. 9, 2012) ("GenProducts is a leading biotechnology company engaging in researching, developing, manufacturing and marketing genomic, proteomic, metabolomic, pharmacologic, bioinformatics, systems-biologic, and healthy products and services.").

28. Caroline McCubbin, *Legal Issues in Biometrics*, 9 J. Comm. Biotech. 249–65 (2003); *see also* DNA Data Bank of Japan, *supra* note 18 ("DDBJ data set is a collaborative work of tens of thousands of authors world wide").

29. *Id.*

30. HHS.gov, http://www.hhs.gov/pia (last visited Nov. 9, 2012) ("Does the system collect PII? No."). "If you are submitting human sequences to GenBank, do not include any data that could reveal the personal identity of the source. GenBank assumes that the submitter has received any necessary informed consent authorizations required prior to submitting sequences." NCBI GenBank, http://www.ncbi.nlm.nih.gov/genbank (last visited Nov. 9, 2012).

31. The EBI website notes in listing projects that "clinical data has been withheld . . . due to privacy concerns." http://www.ebi.ac.uk/ebisearch.ebi?requestFrom=searchBox&db=allebi&query=privacy. In addition, instructions state: "Submit any data that has been consented to for use in research and sharing to appropriate researchers, but has not been consented for full, open public discussion," thereby recognizing the concept of privacy. *See* EMBL-EBI, *Submissions to the EBI Databases*, http://www.ebi.ac.uk/Submissions (last visited Nov. 9, 2012). Moreover, the INSD has a "uniform policy of free and unrestricted access to all data records their databases contain. Scientists worldwide can access these records to plan experiments or publish any analysis or critique." DNA Data Bank of Japan, *supra* note 18.

32. Sethi & Theodos, *supra* note 15.

33. Goodman, *supra* note 13.

34. *Id.*

35. National Academy of Sciences, *supra* note 12.

36. *See* Association of Cancer Online Resources, http://www2.acor.org (last visited Oct. 28, 2012).

37. Guidelines issued by the American College of Medical Genetics in 2007 state: "The testing laboratory . . . should make an effort to contact physicians of previously tested patients in the event that new information changes the initial clinical interpretation of their sequence variant."

38. *See* Luigi Berlinguer & Philippe Lamberts, *Horizon 2020: Inventing in the Common Good, available at* http://www.eagha.org/assets/files/Conference_programme_Horizon_2020_-_Investing_in_the_common_good.pdf.

39. *See* Horizon 2020, http://ec.europa.eu/research/horizon2020/index_en.cfm?pg=excellent-science.

40. 45 C.F.R. pts. 160 & 164. A useful summary with citations can be found at http://www.hhs.gov/ocr/privacy/hipaa/understanding/summary/privacysummary.pdf.

41. 45 C.F.R. § 160.103.

42. *Id.* A covered entity is a health care provider, a health care plan, or a health care clearinghouse, including entities that process nonstandard health information they receive from another entity into a standard format (i.e., standard electronic format or data content), or vice versa.

43. *See* HHS Summary of the HIPAA Privacy Rule, *available at* http://www.hhs.gov/ocr/privacy/.

44. 45 C.F.R. § 106.103.

45. *See* U.S. Department of Health & Human Services, Health Information Privacy (2003), http://www.hhs.gov/ocr/privacy/hipaa/understanding/coveredentities/research.html.

46. 45 C.F.R. § 160.30; NIH Educational Materials, Clinical Research, *available at* http://privacyruleandresearch.nih.gov/pr_06.asp.

47. Business associates perform functions such as claims processing or administration; data analysis, processing, or administration; utilization review; quality assurance; billing; benefit management; practice management; and re-pricing. 45 C.F.R. § 160.103. HIPAA business associate provisions can be found at 45 C.F.R. §§ 164.502(e) & 164.504(e) and are discussed fully in NIH Educational Materials, *Clinical Research*, *available at* http://privacyruleand research.nih.gov/pr_06.asp.

48. 45 C.F.R. § 164.514(a)–(b).

49. *See*, NIH Educational Materials, *supra* note 47.

50. *Id.*

51. 45 C.F.R. pt. 2.

52. 45 C.F.R. pt. 46; 21 C.F.R. pts. 50 & 56.

53. 45 C.F.R. pt. 160 & 164(A)–(C).

54. *See* NIH Educational Materials, *supra* note 47.

55. The Omnibus Final Rule issued by HHS on January 25, 2013, 78 Fed. Reg. 5566, combining all regulatory guidance for HIPAA, HITECH, and GINA together, amends the regulation § 164.508(B)(3)(i) and (iii) to generally allow a covered entity to combine conditioned and unconditioned authorizations for research provided that the authorization clearly permits the individual the option of opting into the unconditioned research. The agency explained that the amendment would permit researchers to combine these authorizations to facilitate research. 78 Fed. Reg. 5610. For example, participation in a clinical trial is often conditioned on the individual's willingness to permit his or her PHI to be use for the research purpose of the clinical trial, i.e., the conditioned authorization. The trial may also include a related component, in the commentary's example, "biospecimen banking," i.e., the unconditioned authorization. A covered entity may not place both authorizations in a single form in order to avoid confusion to research subjects and promote administrative efficiency. In addition, the new regulations also permit authorizations that would permit the individual's PHI to be used in future research so long as the authorization makes clear that the PHI may be used for future research without further authorization. 78 Fed. Reg. 5612; *see also* 45 CFR § 164.508(c)(1)(iv).

56. The 18 identifiers include name, geographic information, all elements of dates, telephone numbers, fax numbers, social security numbers, medical record numbers, health plan beneficiary numbers, account numbers, certificate/license numbers, vehicle identifiers, serial numbers, web universal resource locators, IP addresses, biometric identifiers, photographs and images, and any other unique identifier.

57. 45 C.F.R. § 514(e)(2).

58. Factors include minimal risk to the privacy of individual(s); research cannot be practically conducted without a waiver; and research cannot be practically conducted without access to PHI. 45 C.F.R. § 164.512(i).

59. 45 C.F.R. § 164.512(i).

60. 45 C.F.R. § 164.512(i)(1)(ii).

61. 45 C.F.R. § 164.514(d)(3)(iii)(D).

62. 45 C.F.R. § 164.528(b)(4)(ii).

63. Rothstein, *Is Deidentification Sufficient to Protect Health Privacy in Research?*, NIH PUBLIC ACCESS, AUTHOR MANUSCRIPTS (Feb. 2, 2011) *available at* http://www.ncbi.nlm.nih.gov/pmc/articles/PMC3032399.

64. *See CIPP Guide: De-Identification & Re-Identification*, http://www.cippguide.org/2010/09/21/de-identification-re-identification (last visited Nov. 5, 2012) (citing examples of re-identification).

65. *Id.*

66. *Id.*

67. Bradley Malin & Latanya Sweeney, *Re-identification of DNA through an Automated Linkage Process*, AMIA ANNUAL SYMPOSIUM PROCEEDINGS ARCHIVE, *available at* http://www.ncbi.nlm.nih.gov/pmc/articles/PMC2243547; *see also* Cancer Center WIKI, *Access Control Policy: Commentary*, http://wiki.nci.nih.gov/display/TCGA/Access+control+policy.

68. *Id.*

69. Bradley Malin & Latanya Sweeney, *How (Not) to Protect Genomic Data Privacy in a Distributed Network*, J. BIOMED INFO 2004, cited in Malin, *Re-Identification of Familial Database Records*, AMIA Annual Symposium Proceedings Archive, *available at* http://www.ncbi.nlm.nih.gov/pmc/articles/PMC1839550 (reporting the availability of software to reidentify family relations).

70. Bradley Malin, *Re-Identification of Familial Database Records*, AMIA Annual Symposium Proceedings Archive, *available at* http://www.ncbi.nlm.nih.gov/pmc/articles/PMC1839550 (reporting the availability of software to reidentify family relations).

71. In the commercial context, the FTC requires companies to make data reasonably de-identified, commit to not re-identifying it; and, further, requires anyone who receives it to promise not to re-identify it. *See* Peter Swire, *Consumer Data Privacy Codes of Conduct: Recommendations for the Department of Commerce Multi-Stakeholder Process*, SCIENCE PROGRESS, http://scienceprogress.org/2012/consumer-data-privacy-codes-of-conduct (last visited Apr. 11, 2012). Failure to comply would subject the entity to a lawsuit under Section 5 of the FTC Act for an alleged deceptive practice. *See FTC Protecting Consumer Privacy in an Era of Rapid Change: Recommendations for Businesses and Policymakers*, http://www.ftc.gov/opa/2012/03/privacyframework.shtm (last visited Mar. 2012).

72. *See* Bearder v. Minnesota, 806 N.W.2d 766 (Minn. 2011); Norman-Bloodsaw v. Lawrence Berkeley Laboratory, 135 F.3d 1260, 1269 (9th Cir. 1998).

73. *See* National Conference of State Legislatures, *Genetic Privacy Laws*, http://www.ncsl.org/issues-research/health/genetic-privacy-laws.aspx.

74. *See, e.g.*, Anna Lumelsky, *Genetic Testing and Government Regulation: The Growing Significance of Pharmacogenomics*, *available at* http://leda.law.harvard.edu/leda/data/487/Lumelsky.pdf.

75. 42 U.S.C. § 2000ff.

76. *Id.*

77. *Id.* § 2000ff(4).
78. *Id.* § 2000ff(4)(B).
79. *Id.* § 2000ff(7).
80. *Id.* § 2000ff(6).
81. *Id.* § 2000ff-5(C).
82. The Omnibus Final Rule, 78 Fed. Reg. 5566, 5658 *et seq.*, sets forth the modifications to the HIPAA required by GINA—principally prohibiting the disclosure or use of genetic information for underwriting purposes by all health plans that are covered entities under the HIPAA Privacy Act including those not expressly covered by GINA itself. 78 Fed. Reg. 5660; *see also* 45 CFR 164.502(a)(1)(iv). Most provisions in the final regulation define terms as found in the statute.
83. *See* National Institutes of Health, *Research Repositories, Databases, and the HIPAA Privacy Rules*, http://privacyruleandresearch.nih.gov/research_respositories.asp (last visited Oct. 30, 2012).
84. Traditional business associate functions and activities include: claims processing or administration; data analysis, processing, or administration; utilization review; quality assurance; billing; benefit management; practice management; and repricing. 45 C.F.R. § 160.103.
85. 78 Fed. Reg. 5575.
86. *Id.*
87. *Id.*
88. HHS Summary of the HIPAA Security Rule, http://www.hhs.gov/ocr/privacy/hipaa/understanding/srsummary.html.
89. Large genomic databases maintained by the federal government are covered by the provisions of the Federal Information Security Management Act of 2002 (FISMA), 44 U.S.C. § 3541, that requires, inter alia, risk assessments, and the Privacy Act of 1974, 5 U.S.C. § 552a. *See* NIST Special Publication 800-53, Rev. 3, Recommended Security Controls for Federal Information Systems and Organizations (NIST has published Rev. 4 for comment).
90. 74 Fed. Reg. 19,006 (April 27, 2009); *see also* 45 C.F.R. § 164.
91. *See* the section titled "Health Care Data Breaches and Information Security: Addressing Threats and Risks to Patient Data."
92. *See* GenBank Privacy Impact Assessment, http://nih.gov.pia.
93. *See* Table 13.2.
94. HIPAA security rules can be found at 45 C.F.R. § 164.306 (Security standards—general rules); 45 C.F.R. § 164.308 (Administrative safeguards); 45 C.F.R. § 164.310 (Physical safeguards); 45 C.F.R. § 164.312 (Technical safeguards); and 45 C.F.R. § 164.314 (Organizational requirements).
95. 5 U.S.C. § 552a.
96. 44 U.S.C. §§ 1354 *et seq.* A description of the FISMA requirements and NIST guidance can be found at http://csrc.nist.gov/groups/SMA/fisma/index.html.
97. Title XIII of the American Recovery and Reinvestment Act of 2009 (ARRA), Pub. L. No. 111-5, 123 Stat. 115 (2009); 42 U.S.C. § 17921.
98. The HHS website contains a summary of the HITECH statute at http://www.hhs.gov.ocr.privacy/hipaa/administrative/enforcementrule/hitechenforcementifr.html.
99. 42 C.F.R. pts. 412, 413, 422 & 495 (HHS); 16 C.F.R. § 318.4 (FTC).
100. 78. Fed. Reg. 5641.
101. *Id.*
102. *See* National Conference of State Legislatures, *State Security Breach Notification*, http://www.ncsl.org/issues-research/telecom/security-breach-notification-laws.aspx (last visited Aug. 20, 2012); Thomson, *supra* note 5, at 87–114.
103. 42 CFR § 164.308(a)(1)(ii)(A); 78 Fed. Reg. 5566, 5640–44 (January 25, 2013).
104. HIPAA requires administrative, physical, and technical safeguards, 45 C.F.R. § 164.306.
105. Excellent resources for conducting a risk assessment and assessing compliance with the HIPAA Privacy and Security Rules are available at http://www.nist.gov/itl/csd/20111122_hipaa_tools.cfm.
106. *See* Thomson, *supra* note 5, at 57–85.
107. The high costs of responding to data breaches have been well documented. As the notification requirements in the data breach laws become more onerous with increasing numbers of individuals who must be notified, and the liability imposed by courts and administrative agencies for data breaches increases significantly, these costs are likely to rise dramatically.
108. *Emory Healthcare notifies individuals regarding missing data*, http://news.emory.edu/stories/2012/04/ehc_missing_data/campus.html (last viewed Apr. 20, 2012).
109. In January 2012, the Georgia Health Sciences University in Augusta announced a breach in its research medical school when a laptop computer that may have contained some of the patients' personal information was stolen. Health Informatics, http://www.healthcare-informatics.com/news-item/georgia-health-sciences-university-reports-data-breach.

110. Utah Department of Health, Data Breach Update, *available at* http://utahlegislature.granicus.com/MetaViewer.php?view_id=2&clip_id=1423&meta_id=56353.

111. Joseph Goedert, *Big Medicaid Data Breach in South Carolina*, Health Data Management (Apr. 19, 2012, 5:09 PM), http://www.healthdatamanagement.com/news/breach-notification-hipaa-privacy-security-ocr-44343-1.html?ET=healthdatama.

112. Sutter Health, An Important Notice for Patients Regarding Stolen Computer (Nov. 16, 2011), *available at* http://www.sutterhealth.org/noticeforpatients/.

113. ITRC, http://www.idtheftcenter.org (last visited Nov. 5, 2012).

114. ITRC, *Data Breaches: The Insanity Continues* (Jan. 8, 2012, 8:45 AM), http://www.idtheftcenter.org/artman2/publish/lib_survey/Breaches_2009.shtml.

115. The Privacy Rights Clearinghouse reported 17 known insider breaches by June 2012. DataLoss reported 29 insider breaches during the same period. *See Chronology of Data Breaches Security Breaches 2005–Present*, Privacy Rights Clearinghouse (updated Nov. 4, 2012), http://www.privacyrights.org/data-breach/new; http://www.dataloss.db.org/statistics.

116. *Top Causes of Data Breaches—Negligent Insiders?*, http://www.net-security.org/secworld.php?id=12634 (last visited Nov. 5, 2012).

117. CERT Insider Threat Center, http://www.cert.org/insider_threat/.

118. U.S. Attorney's Office, W.D. Pa., Former UPMC Shadyside Hospital Employee Pleads Guilty to HIPAA Violation (June 30, 2011), *available at* http://www.justice.gov/usao/paw/news/2011/2011_june/2011_06_30_03.html.

119. *See Man Stole ID's of Brain-Injured for Fake Tax Returns*, NYDailyNews.com (Apr. 16, 2012), http://articles.nydailynews.com/2012-04-16/news/31351331_1_tax-refunds-identity-theft-tax-returns (48-count indictment alleging grand larceny in the third degree, identity theft in the second degree, offering a false instrument for filing in the first degree, and possession of a forged instrument in the second degree).

120. J. Brandt, *UHC Warns of Breach* (May 2012), *available at* http://www.kivitv.com/news/local/152036715.html.

121. Letter from the Director, Berkeley Heart Lab to the Office of the New Hampshire Attorney General (October 31, 2011), *available at* http://doj.nih.gov/consumer/security-breaches/documents/berkeley-heartlab-20111031.pdf.

122. Josh Shaul, *Steps to Comprehensive Database Security and Compliance*, http://www.us-cert.gov/GFIRST/presentations/2011/Combating_Database_Insider_Threat.pdf. Specific security measures that must be taken to protect personal data include: (1) inventory your databases; (2) classify systems with sensitive data; (3) scan for vulnerabilities and misconfigurations, keep up to date with security patches, enforce strong passwords, and audit configurations and settings; (4) identify privileged users (DBAs); (5) validate access to sensitive data; assign restricted permissions on tables with sensitive information; (6) prioritize and fix what you can; (7) monitor database activity; and (8) encrypt data in transit and at rest using network-level encryption and column-level encryption.

123. St. Joseph Health Reports Potential Data Breach, N. Bay Bus. J., *available at* http://www.northbaybusinessjournal.com/48980/st-joseph-health-reports-potential-data-breach/.

124. The FTC has notified almost 100 organizations that personal information, including sensitive data about customers and/or employees, has been shared from the organizations' computer networks and is available on peer-to-peer (P2P) file-sharing networks to any users of those networks, who could use it to commit identity theft or fraud. Widespread Data Breaches Uncovered by FTC Probe, FTC Warns of Improper Release of Sensitive Consumer Data on P2P File-Sharing Networks, *available at* http://www.ftc.gov/opa/2010/02/p2palert.shtm.

125. *See* NIST Special Publication 800-147, *Guidelines for Security and Privacy in Public Cloud Computing, available at* http://www.nist.gov/itl/csd/cloud-012412.cfm.

126. The SANS Consensus Audit Guidelines (CAG) provide guidance to maximize the impact of government and private-sector security efforts, and identify 20 critical priority controls, most of which can be continuously monitored. Twenty Critical Security Controls for Effective Cyber Defense: Consensus Audit Guidelines, v. 3.1, *available at* http://www.sans.org/critical-security-controls/ (the website provides a wealth of valuable information about the leading information security methodologies and how they relate to each other).

127. NIST Special Publication 800-53, Rev. 3, *Recommended Security Controls for Federal Information Systems and Organizations* (NIST has published Rev. 4 for comment).

128. NIST Special Publication 800-37, Rev. 1, *Guide for Applying the Risk Management Framework to Federal Information Systems, A Security Life Cycle Approach* (February 2010), *available at* http://csrc.nist.gov/publications/nistpubs/800-37-rev1/sp800-37-rev1-final.pdf.

129. NIST Special Publication 800-66, Rev. 1, *An Introductory Resource Guide for Implementing the Health Insurance Portability and Accountability Act (HIPAA) Security Rule* (October 2008), *available at* http://csrc.nist.gov/publications/nistpubs/-Rev1/SP-800-66-Revision1.pdf (summarizing the HIPAA security standards and explaining the security concepts of the HIPAA Security Rule; providing a crosswalk to the security controls in NIST SP 800-53).

CHAPTER 14

DNA Databanks and Informed Consent

Herman T. Tavani and Maria Bottis*

INTRODUCTION

This chapter examines some challenges for the informed-consent process in medical research where DNA databanks are employed. First, we briefly describe the principle of informed consent and show why it is ethically important. We then examine some specific challenges that arise for the traditional informed-consent process in population-wide genetics/genomics research, especially where data-mining techniques are used. Finally, we defend a model of consent based on the notion of a "charitable trust," which we believe can help to preserve the integrity of the consent process in the context of medical and genetic research involving DNA databanks.[1]

THE INFORMED-CONSENT PROCESS IN THE CONTEXT OF MEDICAL RESEARCH

The principle of informed consent, especially as it applies to medical research, has been staunchly defended by many ethicists. According to Onora O'Neill, for example, informed consent is "ethically significant" for participants (i.e., human subjects) in research because it provides both (a) respect for the person; and (b) protection of the person's autonomy.[2] O'Neill also argues that the principle of informed consent is "ethically important" in that it helps individuals to protect themselves against both coercion and deception.[3] If O'Neill is correct, there are good reasons for defending and preserving the principle of informed consent. However, O'Neill also suggests that an adequate consent process for research subjects should be transparent. We question whether O'Neill's transparency requirement for the informed-consent process can be achieved in population-wide, genetic/genomic research involving the use of data-mining technology and DNA databanks. First, however, we briefly examine the origin and evolution of the informed-consent process.

---

*This chapter is adapted from Herman T. Tavani & Maria Bottis, *The Consent Process in Medical Research Involving DNA Databanks: Some Ethical Implications and Challenges*, 40 ACM SIGCAS Comp. & Soc. 11 (2010).

The Principle of Informed Consent

What exactly do we mean by informed consent? According to Ruth Faden et al., it can be defined as "an autonomous authorization" on the part of a "patient or subject."[4] We should note that in the context of medical research, the principle of informed consent originates from a strict, two-part relationship: the physician-patient relationship. Also, it originates from a situation where a medical intervention (typically surgery) is sought to be authorized by the patient. In this sense, consent has been the strongest part of the "informed consent rule."[5] Since informed consent's inception in *Salgo*,[6] the courts have been slowly accepting and adding new "types" of information necessary for a legal medical intervention. These include information about the nature and purpose of the intervention; diagnosis and prognosis of treatment; alternative modes of treatment; the particular percentages of success or failure; possible conflicts of interest;[7] and the possible experimental nature of treatment.[8]

We believe that the same expansion of information must be offered in the setting of another relationship: medical researcher–research subject. Informed consent in the setting of medical research has usually involved patients who are required to take a drug (or a placebo). In this context, the informed subject must know the nature and purpose of the research, as well as its consequences and risks. In other words, consent must be express (or explicit), specific, and documented.

International documents, such as the Oviedo Biomedicine Convention, also require that some additional information be given to the research subject—namely, information about:

- any foreseen potential future uses, including commercial uses of the research results, data, or biological materials;
- the source of funding of the research project; and
- access to the research results in the future.

Article 13 of the Additional Protocol to the Oviedo Biomedical Convention requires that information given to research subjects is in a form that is "comprehensible" to them. It also states that the opinion of the Ethics Committee, which necessarily supervises the research, should be included in this information. Potential participants must also be informed of their right to refuse or withdraw consent at any time, without being subject to any form of discrimination (e.g., to be refused necessary medical care).

It is also important to note that "undue influence concerns," such as misuse of a position of trust, have led to the implementation of some additional safeguards for human subjects who voluntarily participate in research. The Explanatory Report to the Additional Protocol of the Oviedo Biomedical Convention lists three kinds of "undue influence": physical coercion, financial incentives, and misuse of a position of trust. In the United States, the Code of Federal Regulations provides that the circumstances of consent must give the subject sufficient opportunity to consider whether or not to participate and must minimize the possibility of coercion or undue influence.[9] Additionally, the consent agreement may not include any exculpatory language through which the subject or the representative is made to waive any of the subject's legal rights or release the investigator, the sponsor, the institution, or its agents from liability for negligence.

Is it possible to achieve what some ethicists call "valid informed consent"[10] or "fully informed consent"?[11] Sheri Alpert has argued that two conditions are required for establishing "valid" informed consent in the context of collecting medical data: (1) individuals must "know and understand the nature of the information being released," and (2) consenting individuals must be made aware of the party or parties to whom the information about them can be released.[12] Similar criteria for establishing valid informed consent are specified in the Office of Technology Assessment (OTA) Report.[13] According to OTA, valid informed consent, in the context of medical records, is possible only when individuals are "familiar with the data contained in their records, so that they understand what they are consenting to disclose." This, in turn, requires that patients both: (1) have adequate disclosure of information about the data dissemination process, and (2) are able to fully comprehend what they are being told about the procedure or treatment.[14] In the next section, titled "The Opacity of Consent, Data-Mining Tools, and DNA Databanks," we consider the implications of these two challenges for the informed-consent process. First, however, we briefly examine some implications that a position called "genetic exceptionalism" also may have for this process.

The Challenge Posed by Genetic Exceptionalism

Many who advocate for preserving the informed-consent process in genetic research worry about the highly sensitive nature of personal genetic data vis-à-vis ordinary medical data. For example, some bioethicists argue that genetic data needs additional protection because, unlike ordinary medical data, it is: (1) often "predictive" of a patient's future health status (as opposed to merely used to identify a patient's current medical condition); and (2) able to reveal information not simply about the research subject herself, but about her family members as well. This view is typically referred to in the bioethics literature as the thesis of "genetic exceptionalism" (sometimes also referred to as "genetic essentialism"). Those who support this thesis worry that the misuse of a research subject's personal genetic data could result in that subject's experiencing future discrimination and stigmatization. For example, the person might be denied a job or be unable to qualify for health insurance because of information disclosed in her genetic data.

Those who reject the genetic exceptionalism thesis, who we can refer to as "anti-exceptionalists" for purposes of convenience, believe that personal genetic data is not necessarily different in kind from many forms of ordinary medical data.[15] For example, they note that much ordinary medical data can also be predictive of an individual's future health prospects and that this data can also reveal health-related information about the patient's family members. However, we do not need to pursue the debate between exceptionalists and anti-exceptionalists to understand why protecting one's personal genetic data is important (even if it turns out that this data is not categorically different from some kinds of ordinary medical data about an individual).[16] Instead, we next turn our attention to analyzing the processes used to acquire that information and the nature of the databases used to store and manipulate it.

THE OPACITY OF CONSENT, DATA-MINING TOOLS, AND DNA DATABANKS

One challenge for the consent process in contemporary medical and genetic research has to do with a factor that O'Neill calls the "opacity," or nontransparency, of consent. We will see that the consent process in DNA research involving databanks is nontransparent, in part, because of the secondary uses of that information made possible by techniques such as data mining. But we should also note that it can be "opaque," because the consent process is not one that is necessarily transitive in nature—i.e., one's having consented to X does not necessarily imply that one has consented to Y (even when it would seem that

X logically entails Y). For example, consider the following scenario: The research subject, A, consents to procedure B. Further imagine that B entails C (and that A is unaware of C). Does it follow that A consented to C? O'Neill notes that the consenter may not see that B either entails C or is equivalent to C. She points out that someone

> might consent to a medical procedure described in euphemistic and unthreatening ways, yet not see [her]self consenting to another more forthright and equivalent description of that treatment. . . . [She] might consent to chemotherapy, and yet when as a result [she feels] desperately ill and weak may truthfully claim that [she] never consented to anything that would have this effect—even if these very effects were described among the normal effects of the treatment.[17]

O'Neill's examples illustrate how easy it is for someone to consent to a particular medical procedure and yet still claim not to have consented to procedures that either are roughly equivalent to or entailed by the original consent agreement.

Although O'Neill believes that this kind of opacity, or lack of transparency, in the informed-consent process affects the broader spectrum of contemporary medical research, we believe that it is especially problematic in DNA research involving databanks because of the kinds of subsequent uses of personal genetic information that are made possible by data-mining technology. We will next see how the opaqueness problem is aggravated by the use of data mining.

Data Mining and Secondary Uses of Personal Medical Information

What is data mining, and how does it threaten the informed-consent process used in population genomics research? Data mining (sometimes also referred to as Knowledge Discovery in Databases, or KDD) is a data-manipulation technique that uses pattern-matching algorithms to "discover" patterns in data that would otherwise be nonobvious to the data analyst.[18] For example, a data-mining algorithm might reveal that women who read 18th-century German novels have a higher-than-average incidence of breast cancer. So, a new group of individuals based on nonobvious statistical correlations can be generated by data-mining technology. For those persons affected by this "new classification," however, their association with such a group may seem totally arbitrary to them, because they could argue that it has no genuine basis in reality. Yet, the statistical relationships revealed by data mining generate new groups and "new facts," such as the "group of individuals that is likely to have or to contract breast cancer because they read 18th-century German novels."[19] Furthermore, consider that many members of this (newly discovered/created) group will have no idea that they have been identified as members of it, and thus will have little or no recourse when it comes to their right (and expectation) to be able to correct any inaccurate information about them that links them to the "new group."

Another factor involving data mining is that associations such as the one described in the preceding paragraph can result from secondary uses of data originally collected and intended for use in one very specific study or project (e.g., in some research study targeting a particular disease). That data, after it has been aggregated, could be used in subsequent applications, which can, in turn, reveal new statistical associations (affecting an individual) that would not have been possible to predict or anticipate at the time the data was initially collected and used. So, if the research subject consented to having her data collected in the initial study, for one purpose—namely, the objectives of that particular experiment—we can ask whether she also consented to have her data used (i.e., mined) in subsequent applications involving secondary uses of her personal data. This controversy serves to further illustrate O'Neill's point about the opacity of the informed-consent process in contemporary medical research (described in the previous section).

To see how O'Neill's insight applies to our concern about research subjects in the context of genomics research, consider the following question: When an individual has consented to give DNA for use in one context, does it follow that she has also consented that her DNA can be used in subsequent contexts? For example, does it follow that she has consented that her information can be used in constructing new groups of individuals based on statistical correlations generated from confidential data that is sensitive? And does it follow that she has also consented to having this information cross-referenced with other kinds of information, which could result in adverse judgments being made about her (that put her at significant risk for discrimination or stigmatization)? From what we have seen about the ways that data mining can be used to construct new groups from data derived from contexts in which that data was initially given for very different purposes, it would seem that the process of informed consent has become one that, as O'Neill correctly suggests, is far more opaque than transparent. It would also seem that the kind of conditions required for "valid" or "fully informed" consent are extremely difficult, if not impossible, to achieve in cases of genomics studies involving the use of data-mining technology.

An interesting and corollary issue that also arises in this context is: Who owns the personal genetic data that resides in the researcher's databank? In consenting to allow their personal data to be used in a research study, have the human subjects also consented to give away their data or to grant that it can be used for all possible future applications involving the particular databank in which it is stored? In one sense, questions pertaining to the ownership of the personal genetic data that reside in databanks are beyond the scope of this chapter; however, we believe that this is also an important question for research subjects to consider in the consent process.[20]

Thus far, we have considered some implications for individuals who are adversely affected by the use of data mining in connection with DNA databanks. We next consider the potential impact for entire populations (as opposed to the individuals comprising them).

DNA Databanks and Community Consent Involving Populations

We argue that the use of DNA databanks in population-wide research has introduced some key challenges for what can be viewed as the "traditional research paradigm"[21] in the informed-consent process. Initially, one might assume that the act of donating some blood or tissue carries no risk of immediate harm to that consenting individual. Indeed, the act of giving a sample of one's blood, hair, or body tissue for research would seem harmless; in fact, refusing to do so might be interpreted as an "antisocial" act that also violates the principle of "solidarity" underlying DNA research. Consider that without these research "materials," there can be no DNA research at all. However, we believe that the act of donating one's tissue, blood samples, etc., for research involving a DNA databank presents some new challenges for the informed-consent process because of both (1) the special nature of population-wide genomic research projects, and (2) the kinds of technologies involved (e.g., data mining). For one thing, the familiar two-part relationship involving researcher and patient is seriously threatened in these kinds of projects. It also raises an important question: What is the nature of a "population's consent"? And this question, in turn, raises two additional questions:

(1) What is it that a whole population needs to know before consenting?
(2) How can we ensure "fair consent" from everybody involved?

Consider that any harm resulting from "unconsented" and unprincipled research in this area is directed not only against a person, but also against an entire population.

Another challenge has to do with the so-called DNA/gene factor; we are not yet sure what it is that we are providing when we donate our DNA for research. For example, is it information about our "entire self," i.e., our past, present, and future? Is it information about who we are now, at the moment of donation, or is it also a "future diary"? (These concerns are related to, but also distinct from, issues affecting "genetic exceptionalism," described above).

These challenging questions gained worldwide attention when the story of genetic research in Iceland by a private company, deCODE Genetics, Inc., became known. deCODE, founded by Kari Stefansson in 1996, was granted access to all the health records of the entire Icelandic population as well as to the meticulously maintained genealogical information of all Icelandic citizens. Under an agreement with the Icelandic government, the statute that enabled this research, the Health Database Act, presumed the consent of all patients to release their records to deCODE. Individuals were obliged, therefore, to affirmatively opt out if they did not wish for their records to be given to deCODE.[22] This opportunity, of course, was meaningless for children, as well as for already deceased and mentally impaired persons. Presumed consent and the opt-out requirement received an international rejection,[23] while the Icelandic Supreme Court declared the Health Database Act unconstitutional on the grounds that it breached the right to privacy.[24] Following the events involving deCODE, the presumed consent model for genetic research suffered from severe international criticism.

The issues affecting informed consent in genetic research in general, and population-based genomic research in particular (such as in the case of deCODE), introduce challenges for three distinct models of consent:

- presumed consent (where the researcher takes the tissues, presuming that the person from whom they originate wouldn't disagree);
- informed consent (where documented by signing predrafted forms);
- community consent (where the "leader" of a community consents for every member and the members consent, also, for themselves).

We have already examined the informed consent-process in some detail, but we have only briefly described the notion of presumed consent. Although the latter type of consent has been attacked by critics, some have defended the presumed-consent process. For example, Stefansson argues that traditional epidemiological studies have depended on presumed consent, and that the overall quality of health care (in Iceland, at least) has been improved because of it.[25] He also believes that overly stringent informed-consent policies can lessen the likelihood that our children will have the same quality of health care in the future that we have enjoyed to date. So Stefansson responds to his critics by claiming that there are good reasons for defending the principle of presumed consent.

Stefansson also believes community consent is needed in any adequate scheme involving the collection of medical information that affects a large number of people (as, for example, when conducting a major study that involves an entire population such as Iceland). However, Stefansson concedes that when individuals in that community are asked to donate their DNA samples, informed consent is usually necessary. James Anderson and

Kenneth Goodman make a similar point with respect to the "weight" that should be given to informed consent for collecting genetic data (as opposed to other kinds of medical data) in population-based studies.[26] Interestingly, a comparison of eight international population-based genetic databases shows that consent procedures range from presumed consent of the entire eligible population to recruitment of volunteers with informed consent.[27] Can the informed-consent process used in population genomics research be revised in a way that meets the challenges posed by the need for protection at the "group level" or "collective level"? Goodman describes a category of protection that he refers to sometimes as "group confidentiality"[28] and other times as "collective confidentiality."[29] Goodman poses two important questions in connection with this category of protection. For one thing, he asks whether groups per se have a "right to confidentiality." Goodman also asks whether explicit normative policies and laws aimed at protecting group confidentiality would necessarily benefit vulnerable groups and subgroups. For example, he notes that such policies could have the adverse effect of making it even more difficult for members in those groups to avoid certain kinds of group/subgroup stigma. Although Goodman puts forth some important conceptual and normative questions that need to be answered before we can frame and implement a policy aimed at protecting the confidentiality of groups, he does not provide us with explicit details for how such a policy would be framed.[30] We believe that the outlines of an adequate policy can be found in a scheme that some now refer to as the "charitable trust" model for informed consent.

THE CHARITABLE TRUST MODEL AND INFORMED CONSENT

The charitable trust model for informed consent in the context of "biobanks" was first proposed by David Winickoff and Richard Winickoff, who suggested that the legal model of charitable trust, which had been established to enable other kind of legal relationships and gifts, be applied in the case of genomic databanks.[31] In this case, the person donating her tissue for the biobank—i.e., the donor—is the "settler of the trust" and transfers her property interests involving the tissue to the bank. The donor-settler appoints a trustee for this property, and this trustee is the person/organization that engages in the research. As trustees, these agents are responsible for ensuring the protection of the tissue donations, and they bear fiduciary obligations—i.e., obligations that are more strict than normal contractual obligations—to safeguard the trust and to use it for the benefit of a third party, the beneficiary. In the case of genomic databanks, the beneficiary is the general public, as it also is in the case of other traditional or classic charitable trusts.

We believe that the benefits of using the charitable trust model in the context of genetic research significantly outweigh the disadvantages. One advantage is that the very notion of a charitable trust points toward the direction of an "application of altruism"; here, monetary exchanges and the commodification of genetic data, the standard "plague" of genomics research, do not fit. The idea of a trust also reminds us that giving informed consent by a research subject is, indeed, an act of trust, which builds a trust relationship.[32] Additionally, the party conducting the research becomes a trustee, a steward, instead of a potential profit-seeking agency.

Additionally, Winickoff and Winickoff propose that the architect of the trust provides the donor group with an advisory role in the governance of the trust. In this case, the interests of the donor group involved in the research would not simply "vanish" after the tissue has been donated. Instead, the project would be supervised by an appropriate internal review board (IRB) to help ensure that the interests of the donor group are represented.

The role of the IRB in this case is similar in one sense to that found in related benefit-sharing theories, where members of the donor group receive something in return for their donation. This "something" can be either (1) priority of access to newly developed drugs resulting from the research or (2) a percentage (around 2 percent to 5 percent) of the royalties that may accrue from the research (e.g., from patented drugs or diagnostic tests). But the advantage of the charitable trust model, in opposition to the benefit-sharing theories, is that in the first case, the donor group will be able to participate in the very governance of the trust and thus not merely await some distant, although clearly crafted and theoretically sound, benefits, such as the ones mentioned above. In this sense, the donor group, the "community" (e.g., the Icelanders, the Tongans, the families affected by Canavan disease, etc.) will be given an active role throughout the conducting of the research, while the interests of the general public, as the beneficiary, are also taken into account from the very beginning.

Critics of the charitable trust model maintain that it would face substantial challenges in overcoming various financial and organizational difficulties that a trust might typically present, as well as difficulties in resolving potential conflict-of-interest and privacy issues.[33] Certainly, efficiency in running a databank is a major factor in its success, and we often assume that for-profit models are more efficient than others (even in sensitive areas, presumably distanced from a pure market approach, such as medical research). However, we oppose returning to a for-profit model, which would easily reinforce the strong, current privatization trend in genetic research. Alternatively, we argue for the need to insert (legal and financial) features of efficiency into a charitable trust model for a biobank, where necessary. Doing so may help us to achieve a reformulation of the "gift culture" in the biobank settings, safeguarding the ideal of altruism and public benefit, while also enabling a mechanism that does not work against market logic. Moreover, the model not only satisfies the (now) acknowledged benefit-sharing principle (which, simply stated, gives life to the idea that people themselves are as important to research as the tissues they offer), it also ensures accountability.[34] The model also operates under the assumption that the donor group remains throughout the project not only acknowledged, involved, and protected, but this group is also a "co-governor" of the research project. The charitable trust model aims not merely to "include people," but to "include them well."[35] We believe that this model takes us a long way from where we started—i.e., when a person donated the tissue and then, for all practical reasons, just "disappeared."

CONCLUSION

In this chapter, we briefly examined some challenges for the principle of informed consent in the context of current genetic/genomic research in which DNA databanks are used. We began by examining some arguments for why the consent process is ethically important, and we then showed how that process has become increasingly nontransparent or opaque. In particular, we showed how the opacity in the consent process has been exacerbated by the use of data-mining technology in connection with DNA databanks. Finally, we suggested that a consent process modeled along the lines of the charitable trust framework is the most promising scheme for addressing and resolving many of the current challenges posed to the principle of informed consent as it affects population-wide genetic/genomic research.

NOTES

1. In composing this chapter, we have drawn from material in Bottis's 2005 article and several of Tavani's works. Maria Canellopoulou-Bottis, *DeCode Iceland and Genetic Databanks: Where 'Consent' to Generic Research Means Patenting a Nation's Genes*, *in* Proceedings of the Ethicomp 2005 Conference: Looking Back to the Future, Linkoeping, Sweden (2005); Herman Tavani, *Genomic Research and Data-Mining Technology: Implications for Personal Privacy and Informed Consent*, 6 Ethics & Info. Tech., no. 1, 2004 at 15 [hereinafter Tavani 2004]; Herman Tavani, *Environmental Genomics, Data Mining, and Informed Consent*, *in* Ethics, Computing, and Genomics 167 (Herman Tavani ed., 2006) [hereinafter Tavani 2006]; Herman Tavani, Ethics and Technology: Controversies, Questions, and Strategies in Ethical Computing (2011) [hereinafter Tavani 2011].

2. Onora O'Neill, Autonomy and Trust in Bioethics 2 (2002).

3. *Id.* at 97.

4. Ruth Faden et al., A History and Theory of Informed Consent (1986).

5. Jay Katz, *Informed Consent—A Fairy Tale? Law's Vision*, 39 U. Pitt. L. Rev. 137 (1977).

6. Salgo v. Leland Stanford Jr. Univ. Bd. of Trs., 317 P.2d 170 (Cal. Ct. App. 1957).

7. Moore v. Regents of the Univ. of Cal., 793 P.2d 479 (Cal. 1990).

8. Estrada v. Jaques, 321 S.E.2d 240 (N.C. Ct. App. 1984).

9. 45 C.F.R. § 46.116 (2010).

10. Sheri A. Alpert, *Healthcare Information: Access, Confidentiality, and Good Practice*, *in* Ethics, Computing, and Medicine: Informatics and the Transformation of Healthcare 75, 75–101 (Kenneth E. Goodman ed., 1998).

11. O'Neill, *supra* note 2.

12. Alpert, *supra* note 10.

13. U.S. Congress Office of Tech. Assessment, Protecting Privacy in Computerized Medical Information (1993).

14. *Id.* at 70.

15. Tavani 2004, *supra* note 1, at 15; Tavani 2006, *supra* note 1, at 167.

16. For a detailed discussion of the anti-exceptionalist position, see Tavani 2004, *supra* note 1; Tavani 2006, *supra* note 1. For more information on genetic exceptionalism, see O'Neill, *supra* note 2.

17. O'Neill, *supra* note 2, at 43–44.

18. For a more detailed account of some of the controversies surrounding the data-mining process in the context of genomic research, see Tavani 2004, *supra* note 1.

19. Bart Custers uses a similar analogy in which he envisions a correlation drawn between people who drive red cars and people who have colon cancer. Bart Custers, *The Risks of Epidemiological Data Mining*, *in* Ethics, Computing, and Genomics 153, 153–55 (Herman Tavani ed., 2006).

20. For an analysis of this concern, especially as it relates to either the potential sale or transfer of rights to the information contained in these databases, see Tavani 2011, *supra* note 1.

21. *See, e.g.*, Mark A. Rothstein, *Expanding the Ethical Analysis of Biobanks*, 33 J.L. Med. Ethics 89 (2005).

22. Jamaica Potts, *At Least Give the Natives Glass Beads: An Examination of the Bargain Made Between Iceland and deCODE Genetics with Implications for Global Bioprospecting*, 7 Va. J.L. & Tech. 8, *available at* http://www.vjolt.net/archives.php?issue=13 (2002).

23. Bogi Andersen & Einer Arnason, *Iceland's Database Is Ethically Questionable*, 318 Brit. Med. J. 1565, 1565–67 (1999).

24. Renate Gertz, *Is It "Me" or "We"? Genetic Relations and the Meaning of "Personal Data" Under the Data Protection Directive*, 11 Eur. J. of Health L. 231, 231–44 (2004).

25. *Population, Inc.: How Do You Find Disease-Causing Genes? Kari Stefansson Thinks He Knows the Secret*, 3 Tech. Rev. 50 (Apr. 2001).

26. James A. Anderson & Kenneth E. Goodman, Ethics and Information Technology: A Case-Based Approach to Healthcare Systems in Transition (2002).

27. Melissa Austin et al., *Genebanks: A Comparison of Eight Proposed International Genetic Databases*, 6 Community Genetics 37, 37–45 (2003).

28. Kenneth E. Goodman, Ethics, Computing, and Medicine: Informatics and the Transformation of Healthcare 18 (1998).

29. Kenneth E. Goodman, *Ethics, Genomics, and Information Retrieval*, 26 Computers in Biology & Med. 223, 228 (1996).

30. For an analysis of this point from the perspective of what Anton Vedder (2004) calls "categorical privacy," see Tavani 2006, *supra* note 1.

31. David E. Winickoff & Richard N. Winickoff, *The Charitable Trust as a Model for Genomic Biobanks*, 349 New Eng. J. Med. 1180 (2003).

32. Mats G. Hansson, *Building on Relationships of Trust in Biobank Research*, 31 J. Med. Ethics 415, 415–18 (2005).

33. Jeffrey Otten et al., *Correspondence: The Charitable Trust as a Model for Genomic Biobanks*, 350 New Eng. J. Med. 85 (2004).

34. Cori Hayden, *Taking as Giving: Bioscience, Exchange, and the Politics of Benefit-Sharing*, 37 Soc. Stud. of Sci. 729 (2007).

35. Marilyn Strathern, *Accountability . . . and Ethnography*, *in* Audit Cultures: Anthropological Studies in Accountability, Ethics and the Academy 279 (Marilyn Strathern ed., 2000).

CHAPTER 15

Informatics and Public-Health Surveillance

Leslie P. Francis and John G. Francis

Interoperable electronic health records permit forms of surveillance that may be highly advantageous for public and individual health.[1] The vastly expanding trove of information entered by individuals over the Internet provides additional sources of data for health surveillance, in both the public and the private spheres. Other new digital technologies, such as a geographic information system (GIS), add to the richness of available data. In the United States today, surveillance capabilities are being developed by a variety of public and private entities. These capabilities are being deployed within an incomplete patchwork of legal protections that apply to the data they contain.

This chapter surveys the ethical and legal issues raised by the use of informatics techniques in public-health surveillance. It devotes particular attention to the surveillance capabilities being developed by public entities using data drawn from multiple sources, including electronic health records. Public entities engaged in surveillance face significant ethical tensions between openness to the public about the information they possess and the activities they are carrying out, the need to report results of surveillance to the public to achieve public-health goals and improvements in health, and the importance of protecting privacy and confidentiality of the information used in surveillance. Public entities can expect to receive federal or state Freedom of Information Act (FOIA) requests for results of their analyses or even for the data obtained and used in surveillance. Public entities increasingly may also plan to share their data for others to use. At the same time, the comprehensiveness of data sets, the inclusion of novel data sources, and the power of analytic tools pose new challenges for data privacy. This chapter analyzes the ethical adequacy of legal protections for privacy in the case of public-sector surveillance and recommends additional protections.

BACKGROUND: AN OVERVIEW OF DATA USE FOR HEALTH SURVEILLANCE

Federal and state health agencies, health care institutions, insurance companies, Internet and software service providers, and many other entities have access to large sets of health information that may be useful in identifying patterns of disease, disease outbreaks, treatment efficacy, or unanticipated side effects of therapy. The power of large data sets and novel analytic techniques cannot be described with superlatives that exaggerate; it is grand, encompassing, immediate, and very real. Public and private actors, independently and together, are taking advantage of some of these new capabilities and are planning to take advantage of many more.

In the public sector in the United States today, the federal Centers for Disease Control and Prevention (CDC) and multiple state and local health departments are engaged in extensive surveillance activities. Some of this surveillance resembles the traditional reporting of identified cases of disease that has been a staple of public health for decades, but in exponentially more powerful form. Interoperable electronic health records and other electronic sources of information allow the rapid creation of far more comprehensive, replicable, and transferable data sets that can be subjected to novel analytic tools. Other forms of surveillance, notably syndromic surveillance and detection of adverse events associated with drugs or medical devices, do not depend on the prior identification of diagnosed disease. Instead, they apply analytic techniques to identify patterns that may signify bioterrorism, infectious disease outbreaks, or unanticipated safety concerns. All of these surveillance techniques, but syndromic surveillance in particular, raise complex questions about possibilities for individual consent.

Surveillance activities are not limited to the public sphere. Entities—both not-for-profit and for-profit—undertaking a variety of surveillance activities include health care systems, health insurers, disease-specific advocacy groups, academic institutions, and professional societies such as the American College of Cardiology. Blue Cross, the Cleveland Clinic, and other insurers or providers possessing large sets of clinical data have created for-profit subsidiaries to market their data resources, which can be used in surveillance or for other purposes. Vendors of products—drugs, devices, insurance plans, personal health records, electronic health record systems, systems for managing health information exchanges, and more—also have data collection and analytic capabilities that can be used profitably for surveillance. Finally, the Internet, from Google searches to drug purchases—is a vast and varied source of data about health. Although these surveillance activities are not the subject of this chapter, their existence must be taken into account in assessing the ethics of public surveillance activities.

In this explosion of activity, protections under U.S. law vary depending on the entity collecting or using the information, the type of information, or the information use.[2] With respect to records obtained in clinical care, the most important feature of the current legal landscape is the Health Insurance Portability and Accountability Act (HIPAA), the federal statute governing the protection of most of the information gained in patient treatment and payment for it. HIPAA does not apply to information that has been de-identified to its standards.[3] HIPAA, together with its accompanying regulations, requires patient authorization for many uses and disclosures of individually identifiable health information, but permits health care providers to use or disclose information to the extent minimally necessary for treatment, payment, or health care operations, including activities such as quality improvement or infection control.[4] HIPAA also allows uses or disclosures without authorization for public-health activities as required by law.[5] A wide variety of state laws may implement protections in addition to HIPAA, as long as they do not conflict with HIPAA's requirements. Any health information that is in interstate commerce, whether within or outside of the protection of HIPAA—most importantly the vast amount of information that individuals may enter into the Internet, from Google searches to personal health records to social networking sites—is also regulated by the Federal Trade Commission Act's prohibition of unfair or deceptive trade practices. Finally, many states have fair trade practice laws applicable to this information.

Despite the clear benefits of data use, if protections are inadequate or confusing—or even if there are perceptions that protections may fail—the risk is backlash against important gains from data use. In Texas and Minnesota, to take two of the most high-profile examples, public-health authorities were required to destroy millions of blood spots obtained in newborn screening—an invaluable public-health resource—in response to public outcries about their use.[6] Although these were biological samples, much of the concern about them rested on the information they contained. Understanding available protections and risks of the use of large data sets in surveillance, in both public and private sectors, is thus a matter of urgency if data are to be reliably available for important public-health uses.

PRIVACY, CONFIDENTIALITY, AND CONSENT

A vast literature amplifies the many privacy and confidentiality concerns attending the collection and use of data for surveillance. "Privacy" and "confidentiality" are often conflated in this literature and in the law itself. Informational "privacy" refers to the extent to which health (or other) information is gleaned in the first place.[7] "Confidentiality" refers to control over uses and disclosures of information, once obtained, and "security" refers to means for assuring adherence to specified data protections.[8] Depending on the context, protecting privacy may be the means to protect confidentiality, or the reverse. For example, the response to a confidentiality concern that data transferred for surveillance purposes may later be revealed may be privacy-protective behavior that avoids revealing the information in the first place, as when patients do not seek medical treatment out of concern that their medical records will not be kept confidential. Or, protecting confidentiality may be a means to protecting privacy, as when patients do not want their records shared because they fear that they will be compelled to undergo medical testing or treatment that they do not want.

Values protected by privacy and confidentiality include physical security, liberty, autonomy, intimacy, dignity, identity, and equality. From lepers to persons with human immunodeficiency virus (HIV), victims of infectious disease have been threatened, quarantined, or even killed. The ability to share knowledge of health information may be important to the establishment of intimacy. Health information may reveal aspects of identity such as genetic relationships, ancestry, or disease status. Job loss, denials of insurance coverage, or refusals to provide credit may be the dire economic costs when health information comes to light.

The gravity of these ethical concerns has been judged by some privacy advocates to require informed consent for any collection or use of health information. Leaving aside important questions about whether consent can be adequately informed or effectively protective, strong consent requirements would as a practical matter preclude much use of health information for surveillance purposes. Particularly with the power of large data sets and the development of new analytic capacities, it may be very difficult to anticipate in advance what information might be relevant, how that information might be used, or what results might be expected. Syndromic surveillance is a particular example of this problem. In such cases, patient consent, except of the most general kind—for example, consent to use data in surveillance—is not available to allay concerns about privacy or confidentiality.[9] When data sets involve literally over a hundred million patients, the costs of obtaining individual consent from each of them would be prohibitive. Moreover, many patients may have died, may be without easily identified personal representatives, or simply may have been lost to further contact. If it is to be feasible to use large data sets, such as those drawn from clinical records, for surveillance, then individual informed consent must be judged a problematic method for confidentiality protection.

Some forms of surveillance attempt to avoid privacy and confidentiality concerns by employing so-called de-identified data sets, data sets from which it is believed that the possibility of identifying individuals is "very small."[10] This strategy is given support by data protection regimes such as HIPAA that protect identifiable information but leave de-identified data outside their scope. The strategy of de-identification is not a fully adequate response to confidentiality problems, however. One source of controversy is that, depending on the context and other available information, the risks of re-identification may be greater than generally believed.[11] Moreover, even if individuals cannot be re-identified from the data, it may be possible to draw inferences about them from other known characteristics, particularly if data are analyzed at a sufficiently granular level.[12] For example, if it is verified that everyone with a particular diagnosis (say, cervical cancer) is likely to have been infected with human papillomavirus (HPV), then inferences might be drawn about the sexual histories of persons with cervical cancer. Another set of objections lies in the judgment that individuals should have the opportunity to consent even to uses of "their" data in de-identified form, because they may find certain uses objectionable or because any such uses without consent violate autonomy.[13] Still other concerns address group harms, such as the much-cited example of the use of genetic information obtained from members of the Havasupai tribe for diabetes research, which was later reused to understand migration patterns.[14]

In addition, de-identification may not be a viable strategy because individual identifiers remain critical for public-health purposes. Even in its new forms using large data sets, public-health informatics may need data that have not been fully de-identified. Validation of algorithms for detecting adverse events, for example, may require further information about individuals.

In addition to validating signals, public health may have critical needs to "reach back" to individuals to give them actionable information about their health or the health of others with whom they have been in contact. Such reaching back may benefit the individuals themselves: explanations for symptoms that have not previously been understood, treatment for continuing infection, the recognition of past infection and thus potential subsequent immunity, discontinuation of a drug with likely side effects, or correction or removal of a defective medical device. Other reaching back protects third parties, sometimes but not always with risks to the individuals whose data are the initial source of information: for example, tracing contacts in cases of contagious disease or locating individuals who have received a device or drug to ascertain whether they may also be at risk. In maintaining these reach-back capabilities, syndromic surveillance and surveillance for adverse event detection, even though they begin by searching for unusual patterns of events in data, may—more than is generally recognized—resemble traditional public-health surveillance that relies on case reports of identifiable conditions.[15]

In justifying confidentiality protection, it is important to emphasize that these surveillance activities involve the collection and use of information that was originally obtained for other purposes. These are called "secondary uses" and, because they are by definition different from the original use, they may not be anticipated and thus may surprise or offend individuals or groups.[16] Concerns about secondary uses of health information are particularly strong when the information was originally obtained within the trust relationship of clinical care. Electronic health records (EHRs) are increasingly a source of information for surveillance of the types described below. However, even when individuals enter the far more porous world of the Internet, expectations about information use may matter to them.[17]

Consequentialist justifications for surveillance emphasize the overall aggregate benefit to be achieved. These justifications may fail to protect especially strong individual interests or to attend to considerations of nondiscrimination or social justice. Alan Rubel, for example, defends the view that public-health surveillance activities are not justified in obtaining information without individual consent when claims to privacy are based on "deep personal interests" on the part of individuals in ordering their lives and where exercising this claim to privacy would not be likely to cause others serious injury, illness, or death.[18]

When governmental entities use information in surveillance, further ethical requirements may apply. Openness is one: members of the public may want to know what information is being collected about them and how it is being used by their governments. People also may believe that governments have special responsibilities to share important findings with the public. Such expectations of "sunshine" on the part of government place additional pressures on protection of privacy and confidentiality, as described in the next section.

GOVERNMENTAL OPENNESS AND INFORMATION PROTECTION: LEGAL STANDARDS

Governments in the United States—local, state, and federal—are expected to keep citizens informed. With the possible exceptions of law enforcement and national security—and even here the justification and scope of the exceptions are controversial—governmental data collection activities should be public.[19] Moreover, governments should keep citizens abreast of important developments affecting public safety. In furtherance of such transparency, and with the aim of improving community health, the federal government has encouraged release for public use of some of its vast stores of data.[20] These expectations are not imposed, at least in the same way, on actors in the private sector.

Government data collection and use are circumscribed both by laws requiring openness and by laws protecting privacy and confidentiality. This section summarizes legal protections for privacy and confidentiality of data in the possession of U.S. federal and state governments, together with legal requirements for disclosure. For the federal government, the Privacy Act and the FOIA provide twinned protections for privacy and openness. Special statutory protections for data confidentiality also apply to the Census Bureau, the National Center for Health Statistics, and statistical information collected by other federal agencies under promises of confidentiality. State governments also apply a variety of confidentiality and open disclosure laws to information they collect.

U.S. Privacy Act

The federal Privacy Act originated from a report published in 1973 by the then-Department of Health, Education, and Welfare. The report developed what have become known as Fair Information Practice Principles (FIPPs) to apply to data collection and use made possible by the invention of computers.[21] Although the report recommended universal legal protections, it resulted in enactment of the far-narrower Privacy Act, which applies only to records possessed by federal agencies.[22] With limited exceptions, the Privacy Act prohibits disclosure of identifying information held about citizens or lawful permanent residents without their consent. Included in this prohibition are disclosures to other federal agencies, such as disclosure of health information from the Social Security Administration to the Federal Aviation Administration for use in judging qualifications for pilot licensure.[23] The prohibition does not apply to statistical analyses reported in aggregate form that could not be used to identify individuals or to affect individuals' rights, benefits, or privileges.

The Privacy Act exceptions of greatest relevance to this chapter are (b)(2), the exception for disclosures required by the federal FOIA, described in the following subsection, and (b)(8), the health or safety exception, permitting disclosure "to a person pursuant to a showing of compelling circumstances affecting the health or safety of an individual if upon such disclosure notification is transmitted to the last known address of such individual."[24] For most disclosures, agencies must make reasonable efforts prior to disclosure to assure that records are "accurate, complete, timely, and relevant for agency purposes."[25] This requirement does not apply to (b)(2) FOIA disclosures; agencies may respond to FOIA requests without prior efforts to ascertain information accuracy.

In addition to restricting disclosures, the Privacy Act requires federal agencies to allow individuals upon request to inspect records held about them and to request correction of errors.[26] Agencies may collect only that amount of identifying information relevant and necessary to accomplish their legally mandated purposes.[27] Agencies also must publish in the Federal Register notice about systems of records they collect.[28]

These Privacy Act requirements apply to any collection of identifiable information about citizens or lawful permanent residents by federal agencies such as the Department of Health and Human Services (HHS) or the Food and Drug Administration (FDA). Surveillance activities involving information that could identify these individuals would come under these protections. This would include case reports, adverse event reports, or any health records with identifying information. Statistical analyses using aggregate data, such as general reports of disease incidence or prevalence, would not trigger Privacy Act protections, as they do not include identifiable information. Reports containing health information about persons who have died or noncitizens who do not have the status of lawful permanent residents in the United States also would not receive Privacy Act protection.

Freedom of Information Act

The federal Freedom of Information Act (FOIA) works in tandem with the Privacy Act, which amended it. First enacted in 1966, FOIA requires federal agencies to make information in their possession available to the public.[29] Agencies must respond to FOIA requests that "reasonably" describe the records sought and that are in accord with agency rules.[30]

Several important exceptions to the FOIA disclosure requirement are relevant to this chapter. Agencies may refuse to disclose information that is covered by one of these exceptions. Exception (b)(3) protects information specifically protected by another federal statute (other than the federal Open Meetings Act, the companion statute to FOIA) if the statute "requires that the matters be withheld from the public in such a manner as to leave no discretion on the issue" or "establishes particular criteria for withholding or refers to particular types of matters to be withheld."[31] Statutory protections enacted after 2009 must refer specifically to this FOIA exception.[32] Exception (b)(4) protects from FOIA disclosure "trade secrets and commercial or financial information obtained from a person and privileged or confidential."[33] Exception (b)(6) protects from FOIA disclosure "personnel and medical files and similar files the disclosure of which would constitute a clearly unwarranted invasion of personal privacy."[34] Records that can reasonably be segregated must be provided with redaction of the protected material.[35]

FOIA requests for information possessed by the federal government may be made by any member of the public and requests for health information may be expected from many quarters. Journalists reporting the adequacy of federal government action, personal injury lawyers litigating claims against health care providers or manufacturers of health care products, consumer groups investigating interests of those they represent, and academics researching how laws and agencies function are especially likely sources of FOIA requests. To take one example, FOIA requests for aggregate data concerning the treatment of wounded veterans resulted in a series of stories critical of the Veteran's Administration.[36] As discussed above, the Privacy Act does not require federal agencies to vet the accuracy or relevance of data before responding to FOIA requests.

Upon receiving FOIA requests, agencies must decide whether to invoke FOIA exceptions. Requesters may challenge agency decisions to claim an exception in federal court. Also, individuals may bring "reverse FOIA" suits challenging any public disclosures of information by agencies that are alleged to violate the protections of the Privacy Act.[37] Reverse FOIA suits may be brought to challenge disclosures regardless of whether a FOIA request has been made.

Information used for surveillance and in the possession of federal agencies such as CDC or HHS thus may be the subject of FOIA requests and disclosure would be required unless it comes within a FOIA exception. The Privacy Act specifically does not block disclosures required by FOIA, so it cannot be used to claim a FOIA exception. Thus agencies responding to FOIA requests for information collected for surveillance may assert that it is specifically protected by a statute other than the Privacy Act (for example, the Public Health Service Act), that it is protected commercial information (for example, information that might give providers a competitive advantage), or that it reveals medical files, the disclosure of which would "constitute a clearly unwarranted invasion" of personal privacy. The discussion in the next section of selected surveillance activities by federal agencies will consider how these exceptions might apply to each case. Two important developments in contemporary FOIA law suggest that privacy protection may not be straightforward or easy, however.

First, the U.S. Supreme Court has recently given a restrictive interpretation of another FOIA exception, the exception for records concerning the internal personnel rules and practices of an agency.[38] The decision reasoned that because FOIA is a "disclosure" rather than a "withholding" statute, any exceptions to disclosure must be narrowly construed.[39] The Court's recent wide-ranging free speech jurisprudence also may support readings favoring information disclosure.[40]

Second, in determining whether information disclosure comes within the (b)(6) exception for medical and other private records, courts have applied a balancing test. The test determines whether significant privacy interests are implicated by the disclosure and, if so, whether the public interest in the information outweighs the individual privacy interest. Courts have applied this balancing to medical information, as illustrated by disclosure of a summary of a truck driver's medical records in connection with an investigation by the National Transportation Safety Board (NTSB). Creed was a truck driver who fell asleep at the wheel, causing a massive, multi-fatality accident. The NTSB investigated whether the public safety required new rules governing trucker's hours and medical conditions.[41] When information drawn from Creed's medical records including diagnoses of sleep apnea and hypertension was posted on the NTSB website, Creed brought a reverse FOIA action claiming the disclosure violated the Privacy Act. Creed's primary concern was that in ongoing litigation concerning the accident, the court had not allowed plaintiffs access to his medical records. The NTSB removed the record summary from its website, but allowed it to be discussed in a public meeting over Creed's objection. NTSB's position was that the public interest in understanding the cause of the accident and in making safety recommendations outweighed the trucker's privacy interest. Under the jurisdictional statute for the NTSB, the case was transferred to the U.S. Court of Appeals for the D.C. Circuit, where it was later dismissed on motion from the NTSB.[42] The information from the trucker's medical record remains available to the public on the NTSB website as of this writing.[43]

Applying the balancing framework in *Creed* to information concerning the safety of medical care or medical products would require first determining whether any individual privacy interests are implicated, as they would be by disclosure of identifiable health information. These individual privacy interests would then be weighed against public interests such as health care safety. If the information were sufficiently relevant to determining how to improve patient safety, and if the need for improvement were sufficiently strong, interests in privacy of the medical record arguably could be overridden. Indeed, experience with disclosures under the federal Vaccine Injury Compensation Program (VICP) described below suggests exactly this analysis.[44]

The controversy about possible links between vaccination and autism has prompted advocates to use FOIA requests to investigate whether the VICP has been covering up cases that might demonstrate a relationship.[45] In the important *Long/Burnham* case, advocates at Syracuse University's Transactional Records Access Clearinghouse made a FOIA request for an electronic copy of all records contained in the CASES database of the Civil Division of the Department of Justice (DOJ),[46] including all filings under the VICP. An initial court ruling concluded that the request did not come within the FOIA exception for information protected by statute because the statutory language of the VICP did not protect date and type of vaccine administration as identifying information.[47] On motion for reconsideration, the district court held that this initial decision had relied on the wrong section of the VICP Act[48] and that the applicable section did indeed protect the information.[49] This analysis relies on the statutory structure of the VICP Act, however, and different statutory language could support a conclusion that the information did not come within the statutory protection claimed. Indeed, other courts have held the special master's ruling on VICP compensation, including medical information germane to the decision about whether compensation was required, must be publicly reported under yet a different section of the VICP Act unless it is protected commercial information[50] or medical files, the release of which would be a clearly unwarranted invasion of privacy.[51]

In *Long/Burnham*, the DOJ also contended that types of vaccines administered and dates of administration were protected under the FOIA exception for private information because under the database structure the case caption, which contained the name of the patient (or the person acting on the patient's behalf), would allow linkage of the medical information to individuals. The district court identified an individual privacy interest because the information was "similar to a medical or personnel file" and could be linked to an individual through the CASES docket. It then balanced this privacy interest against the interests of the public in understanding the administration of the VICP funds with respect to the alleged connection between autism and vaccination. In granting summary judgment for the DOJ, the court reasoned that release of the information did not serve a significant public interest because knowledge of the date and type of vaccine would not help to determine whether any compensation claim was autism-related, and because the information was already available in CASES. On this analysis, the balance could be struck differently if the information had been clearly relevant to the autism issue or otherwise unavailable.

Statutes Protecting Government Statistics

As described above, FOIA has an exception for information specifically protected by other federal statutes. Information collected by the U.S. Census Bureau and the National Center for Health Statistics has long enjoyed such special statutory protection; indeed, the history of the Census Bureau, demonstrates that confidentiality was essential to people's willingness to share information with their government.[52] These agencies collect critical national data that may be difficult to obtain even with confidentiality protection. Information collected by the Census Bureau may not be used for any purposes other than the statistical purposes for which it is obtained, published in a manner in which individuals can be identified, shared with other government agencies, or furnished in response to a legal process.[53] In addition to the Census Bureau, the Secretary of HHS has statutory authority to conduct and support health statistics projects through the Agency for Health Care Policy and Research and the National Center for Health Statistics.[54] Any identifiable data thus collected may be used only for the purpose for which it was supplied and may not be published in a manner that would reveal identifiable information, without the consent of the individuals concerned.[55] This statutory section has been applied by the U.S. Court of Appeals for the Fifth Circuit to bar release to Congress of identifiable cause-of-death data obtained by NCHS for research on environmental hazards. In this case, the Fifth Circuit also determined that the contract between NCHS and states for release of the data did not show intent to restrict access of congressional committees to the data but did show intent to avoid outside release of information identifying specific individuals and their causes of death.[56]

Most recently, the 2002 e-Government Act attempts to improve coordination and efficiency of government data collection.[57] Although enhancing data use and minimizing burdens of data collection for the private sector were primary goals of the legislation, the e-Government Act also required oversight of data privacy and security policies within the federal government[58] and development of data privacy and security policies by federal agencies.[59] The Confidential Information Protection and Statistical Efficiency Act (CIPSEA), part of the e-Government Act, expanded the protection of statistical information beyond the Census Bureau and the National Center for Health Statistics[60] to protect statistical information collected by any federal agency under what is called an assurance of confidentiality.[61]

Certificates of Confidentiality for Research Data

In addition to the protections for federal statistics, the secretary of HHS has general authority under the Public Health Service Act to support research and to authorize protection of the confidentiality of research subjects in a manner that withholds any identifying information and protects such information from a compelled legal process.[62] The secretary has delegated this authority to the National Institutes of Health (NIH) to issue certificates of confidentiality granting such protection.[63] This mechanism may provide confidentiality protection to data collected in research, whether or not it is federally funded.[64] When research data is covered by a certificate, investigators subject to a legal process must decide whether to reveal the information or to resist the effort to obtain it. However, any informed consent form must tell research subjects what voluntary disclosures the investigator may make. Informed-consent forms thus frequently specify that researchers will reveal information about suspected child abuse, violence, contagious disease, and the like. Certificates of confidentiality are also limited in that they do not cover information that is not considered sensitive or information that is de-identified. They also do not protect against required disclosures to the FDA. Finally, although these certificates have been in existence for some years, they have yet to be subjected to a definitive test in court. One very early case (in 1973) apparently upheld a certificate as protecting against disclosure in a murder trial of records involving drug treatment research, but another more recent case[65] involving appeal of a conviction for child sexual abuse may have put certificates' authority into question.[66]

State Law

Each state has its own FOIA, applying to data held by its governmental entities. HIPAA does not require patient authorization for use or disclosure of identifiable health records for public health as required by state law, and confidentiality protections for this data are for states to determine. Whether information is subject to disclosure thus depends on privacy and disclosure laws of the individual state in question.

Although frequently modeled on the federal FOIA, these state laws vary in their requirements and restrictions and in how state courts have interpreted them.[67] These state laws may give protections that are stronger than the federal laws, but they also may give more limited protection. For example, states may have language that is broader than the federal (b)(6) exception for "personnel and medical files and similar files the disclosure of which would constitute a clearly unwarranted invasion of personal privacy."[68] Or, states may strike different balances between privacy interests and the public's need to know in such cases. Some state FOIA litigation has involved highly contested political issues; as health care and health issues remain political hot buttons, this litigation can be expected to continue.

Several states have faced lawsuits seeking access to information about abortions reported to the state health department. In Arkansas, for example, a Christian action group brought suit when its request for copies of the reports was denied. The filed forms identified the facility and contained information about the patient such as age, marital status, ZIP Code, date of last menses, and race, although not the patient's name or exact address. The trial court ruled in favor of disclosure, but the Arkansas Supreme Court determined the forms were vital records within the statutory definition and hence exempt from disclosure under Arkansas' FOIA exception for vital records.[69] Had the statutory definition included only medical records but not vital records, however, the information would have been publicly available. Kansas is another state with a conflicted history of public officials and private entities seeking access to abortion records; at present, both the Kansas law requiring abortion reporting and the Kansas Open Records law are interpreted to prohibit disclosure of information identifying either patients or clinics but to permit disclosure of more general information about abortions performed within the state.[70] Public hospitals in other states have also been required to disclose general statistical information about abortions and abortion counseling practices, so long as the information does not reveal patient names.[71] In these cases, courts have apparently assumed that privacy is protected if names of patients (or in some cases facilities) are not revealed, but they have not analyzed the possibilities that the statistical information thus revealed could be used to identify patients.

In some states, requests for disclosure of sick leave records of state employees have been granted, even though these are personnel records and might reveal at least general information about the employee's health status. For example, in Connecticut a taxpayer watchdog group sought disclosure of sick leave records of a school psychologist. Because the request sought information about the number of sick days she had taken, the amount of pay she had received, and her expected date of return, but not directly about her medical condition, the Connecticut Supreme Court concluded that it did not fall within Connecticut's exemption for personnel records, the disclosure of which would be an unwarranted interference with personal privacy. To analyze the concept of an unwarranted invasion of privacy, the court turned to the tort law standard and concluded that revelation of the information would not be highly offensive to a reasonable person.[72] The psychologist had argued that revealing the information would indicate whether she was "sickly" or "healthy," but the court found this reasoning conclusory.

Investigative reporting has also sought information about disease clusters from state cancer registries. In *Southern Illinoisan v. Illinois Department of Health*, a FOIA request for cancer registry data by ZIP Code, date of diagnosis, and type of cancer was filed by an Illinois newspaper. The state contended that the information could identify individuals and refused to release it. Despite the state's expert testimony demonstrating the possibility of individual identification, the Illinois Supreme Court held in favor of the newspaper, reasoning that it was unlikely that members of the public would be able to identify the individuals concerned.[73]

All of these state court decisions do protect information identifying patients by name. However, courts appear to have limited understanding of the powers of contemporary data analytics. State FOIA statutes, like the federal statute drafted during the 1960s and 1970s, contain a patchwork of exceptions that may not apply to all of the current types of data collection and analysis being undertaken by states, such as all payer claims databases or health information exchanges. These data may be of high political interest, as the case cited above illustrates. Whether the data are adequately protected under state law will be a matter of ongoing concern.

The next sections of this chapter describe several primary examples of surveillance for the public good involving federal and state governments and using large sets of health data. It considers the database structure, the extent to which identifiable information is shared or reach-back options are maintained, and the extent of federal or state legal protections in these circumstances.

SURVEILLANCE: NEDSS, BIOSENSE, AND THE NATIONAL HEALTHCARE SAFETY NETWORK

In the United States, public-health agencies now use information from clinical records as well as many other sources for surveillance at national, state, and local levels. The standard-setting umbrella is the Public Health Information Network (PHIN) of the CDC.[74] PHIN coordinates disease-specific and syndromic surveillance activities of the federal and state governments such as the National Electronic Disease Surveillance System (NEDSS) and BioSense. The National Healthcare Safety Network (NHSN) is a separate program sponsored by the CDC Division of Healthcare Quality for monitoring health care safety issues including nosocomial and blood-borne infections. These are among the most important surveillance activities undertaken by CDC, and they illustrate the data-protection issues involved in public-sector surveillance.

The National Electronic Disease Surveillance System

The National Electronic Disease Surveillance System (NEDSS) is the general system for disease-specific surveillance in the United States—that is, for monitoring the incidence and prevalence of diseases of interest. NEDSS is a system of applications under ongoing development by CDC together with state and local health department partners and commercial services and vendors.[75] The goal of NEDSS is to replace the prior patchwork of public-health surveillance systems with a more integrated electronic system.[76] Under NEDSS, CDC receives an extensive amount of identifiable health information, increasingly in electronic form from clinical records.

NEDSS is currently encouraged by the federal government's meaningful-use incentives for health care providers to adopt EHRs.[77] In order to qualify for incentive payments for adopting electronic records, providers must meet specified core standards for the use of these records and a selection of standards from an optional menu.[78] Selected capacities important to surveillance, such as electronic transfer of laboratory results, are included within the meaningful-use optional menu.[79] Despite the meaningful-use incentives and other incentive efforts, progress by NEDSS toward a functional integrated surveillance system remains somewhat halting.[80] Thirty-seven states have the capacity for contact tracing, 34 for case management, 22 for outbreak management, and only seven for receiving electronic health records. NEDSS is thus an extensive, albeit incomplete, system for case reporting to CDC and for sending relevant information back to state health departments.

Information collected through NEDSS receives the federal and state legal protections described above. Federal protections include, most importantly, CIPSEA protection for any statistical information collected under assurances of confidentiality. CDC publishes assurances of confidentiality for each reportable condition.[81] If abstracted data are released for public use, it is with the specification that they may be used for statistical purposes only (for example, analyses of prevalence), contain no identifying information, and not used to try to identify individuals. Federal protections also include the Privacy Act, so CDC must publish notice of any identifiable record systems it maintains and allow individuals the right to inspect and request correction of any records held about them. FOIA requests for identifying information would come under the exception for information protected by statute—in this case, CIPSEA. However, were this statutory exception to be inapplicable (either because CDC failed to provide an assurance of confidentiality or because the assurance of confidentiality structure was altered by Congress or the courts), then the most likely FOIA exception would be for protection of medical records, disclosure of which would be a clearly unwarranted violation of privacy. In such cases, courts would identify the privacy interest and weigh it against the benefits to the public of information disclosure.

Syndromic Surveillance: BioSense

NEDSS is a disease-based surveillance system, relying on the compilation of data about diagnosed cases. Syndromic surveillance, by contrast, identifies patterns that may signal circumstances of interest to public health before any diagnosis has been made.[82]

Experience with syndromic surveillance dates back now over ten years and was stimulated in the United States by the 9/11 terrorist attacks and the anthrax scare.[83] This form of surveillance has been used especially successfully for the identification of influenza outbreaks from respiratory symptoms and for the identification of patterns of gastrointestinal complaints signaling norovirus or rotavirus infection.[84] Patterns of elevated fever in adult patients are thought to be possible indicators of bioterrorism, although there are no reported examples of bioterrorism actually identified in this manner.[85] Typical kinds of data drawn from clinical records for syndromic surveillance are chief complaints reported in emergency department visits and reports of symptoms such as fever or diarrhea in EHRs of ambulatory care providers. Reports of absenteeism rates in schools, purchases of over-the-counter medications in pharmacies, Internet searches, or veterinary medicine visits are also employed. Some of this information—for example, the information in clinical records—originates as HIPAA protected, but other information—for example, Internet search records—does not.

Methods of syndromic surveillance include algorithms for the identification of aberrant patterns. An ongoing problem in the development of these methods is the balance between sensitivity—the goal of identifying (not missing) a pattern of significance—and specificity—the goal of identifying only those patterns that are of actual significance. Too little sensitivity risks the failure to identify events of public-health importance; insufficient specificity may be very costly as it will require many public-health department investigations of false alarms. Studies have suggested that sensitivity and specificity may improve with a focus on local area variations, a finding of particular importance for the protection of confidentiality as it may require highly granular data.[86]

BioSense is the federal syndromic surveillance system for early detection of bioterrorism and other public-health events. BioSense is designed by CDC to provide stakeholders such as state and local health departments and other federal programs with data concerning disease trends that is as close as possible to real time.[87] BioSense, established under enhanced CDC abilities created in the Public Health Security and Bioterrorism Preparedness and Response Act of 2002,[88] has recently been redesigned to be more responsive to the needs of public-health users.[89] Although Biosense reportedly did not originally maintain the capability to reach back to individuals,[90] among the six core data elements included in BioSense redesign are unique facility identifier (such as the national provider identifier) and unique patient identifier, such as patient account number or master patient index number.[91] Entities providing data to BioSense will be able to specify access controls, however.[92]

Like NEDSS, the contemporary version of BioSense also is supported by the "meaningful-use" standards for receiving incentive payments for the adoption of electronic health records. As the meaningful-use standards were developed, an initially proposed implementation specification for the syndromic surveillance objective was withdrawn based on public-health feedback. The CDC then supported an effort by the International Society for Disease Surveillance to recommend requirements for syndromic surveillance practices.[93] These recommendations, issued in 2011 and used by BioSense, include the capability described in the preceding paragraph for reach back to acquire additional patient data when a possible concern of public-health significance has been identified.[94] The capability to report information electronically for syndromic surveillance is an optional menu item in stage 1 meaningful use and is proposed to remain so for stage 2 for eligible ambulatory care providers, out of concerns that health departments are not yet sufficiently equipped to receive ambulatory syndromic surveillance data electronically.[95] It is proposed as a core menu item for eligible hospitals, however.[96]

Any identifiable information collected by CDC for BioSense is obtained under an assurance of confidentiality.[97] Like NEDSS, BioSense is subject to the Privacy Act requirements and FOIA exceptions. States have many requirements for reporting information used in syndromic surveillance and the various confidentiality and disclosure laws described above.

The National Healthcare Safety Network

The National Healthcare Safety Network (NHSN), also maintained by CDC, integrates safety surveillance systems, including efforts to monitor adverse reactions and incidents associated with blood and blood products.[98] NHSN collects data from more than 3,000 hospitals. In 26 states, hospitals are required to report hospital-acquired infections using NHSN;[99] in other states, the reports are made voluntarily. Data reported under NHSN could identify treating facilities and individual patients, and NHSN participants submitting data are provided with an assurance of confidentiality.

In October 2011, NHSN established new capacity for states without mandatory reporting requirements to gain access to data reported by facilities to NHSN, in effect letting NHSN serve as a conduit for transfer of safety information back to the states. Facilities in such states are to be given the opportunity to opt out of voluntary reporting before state access begins. States accessing data are required to enter into a data use agreement (DUA) with CDC, requiring compliance with all applicable state and federal laws, all patient confidentiality and medical record requirements, and any applicable requirements imposed by Institutional Review Boards for the protection of human subjects in research.[100] In addition to protecting individuals who might be identifiable, this DUA is clearly designed to protect the interests of facilities supplying data. It specifies that data may be used only for surveillance (estimating the incidence and prevalence of infections) and prevention (assessing whether a targeted program has reduced infection frequency). Data may not be used for public reporting of institution-specific data or for undertaking regulatory actions against institutions.[101] Thus states accessing data voluntarily submitted by their hospitals to NHSN may not use the data to make public information that might give some facilities competitive advantages over others. These DUA specifications are in addition to any applicable state law protections.

ADVERSE EVENT SURVEILLANCE: MINI-SENTINEL

Concerns about the ability to detect adverse events associated with marketed drugs and devices led Congress in 2007 to mandate additional FDA surveillance activities. Low-frequency or longer-term side effects may pass undetected in the clinical trials submitted for premarketing approval. The FDA Amendments Act (FDAAA) of 2007[102] required the FDA to develop within two years a method for accessing data for active post-marketing surveillance of marketed drugs.[103] The statutory target was to include data from 25 million persons by 2010 and from 100 million by 2012. Congress mandated that the method comply with HIPAA and that it not reveal "individually identifiable health information when presenting such drug safety signals and trends or when responding to inquiries regarding such drug safety signals and trends." Neither of these protections prevents the FDA from developing a system that collects identifiable information without patient authorization or that preserves the capability to reach back to individuals.

The FDA has named the long-term project created in response to the FDAAA mandate "Sentinel" and a shorter-term pilot project "Mini-Sentinel." Mini-Sentinel is composed of a coordinating center and data partners and has developed a governance structure and set of policies for data collection and protection.[104] It is structured so that data remain with the data partners, queries are distributed to the partners, and query responses consist of information in summary form.[105] Mini-Sentinel partners maintain control of all original source data, including claims data, electronic health records, demographic information, and registry data. Data analysis is conducted in standardized form by each partner, using software provided by Mini-Sentinel and a common data model.[106] The summary responses to queries that are shared with the Mini-Sentinel Center contain only the minimum necessary information,[107] stripped of direct identifiers in the HIPAA sense and reported in aggregate form "unless there is specific need for person-level information" such as dates of service. The current need for person-level information is validation of the algorithm used to identify health outcomes of interest—for example, that the algorithm accurately selects patients as having a myocardial infarction and not a different type of health event.[108] However, in the future it is possible that types of information such as ZIP Code below the three-digit level or even more granular geocoding data might be desirable to establish linkages to census data to account for possible environmental factors, or genetic information to determine whether adverse events may be limited to particular genotypes. At present, these uses are not contemplated by Mini-Sentinel. Although they anticipate that any reaching back to contact patients would occur through data partners, Mini-Sentinel planners have not planned to inform patients of the possibility of valid signals of adverse events; instead, any information gleaned from Mini-Sentinel will be provided as a general advisory to physicians or through general publicity.

As thus constituted, Mini-Sentinel appears to have two privacy advantages over NEDSS and NHSN. First, it uses an architecture in which data remain in the hands of data partners, except in the very limited case of data minimally necessary for validation. Second, the FDAAA statute explicitly prohibits disclosure of individually identifying information. Information obtained under the Sentinel program would thus come within the FOIA exception for statutory prohibition of disclosure—but most likely would not even be possessed in the first place.

STATE SURVEILLANCE PRACTICES: A PATCHWORK OF PROTECTIONS

States also collect wide varieties of data for surveillance. These databases often do not have the confidentiality advantages identified above for the federal Mini-Sentinel program. Two types of state databases are described here: the long-standing state cancer registries and the far newer state all payer claims databases (APCDs).

All 50 states and the District of Columbia maintain state cancer registries.[109] These registries can be used by the public to ascertain cancer incidence rates by state and type of cancer and whether these rates are rising or falling. They can also be used to generate micro-maps of cancer death rates by location or by race or ethnicity. Data are reported to the registries as required by state law; patients need not be informed of the reports, and HIPAA authorization is not required because the data are collected under state public-health powers.[110]

State cancer registries are protected by any applicable state laws and are subject to state FOIA requirements. As illustrated by the *Southern Illinoisan* case discussed above, these requirements may vary in strength and may not always protect information that is identifiable, even when the data do not contain obvious links to persons.[111]

In another large-scale data initiative, some states are now collecting APCDs. Nine states currently have established APCDs and many others are actively exploring the possibility.[112] These databases assemble claims data that can be linked for individual episodes of care. For a patient with diabetes, for example, an APCD will link all costs of care attributable to diabetes over a year: hospitalizations, prescriptions, physician visits, test kits, and so on. These databases thus enable comparisons of cost—and potentially of quality—on an episode-of-care basis rather than for particular services. They also enable comparisons of patterns of care by provider, location, or facility. Efforts are being made by the National Association of Health Data Organizations to standardize APCDs to allow comparisons among states.[113]

These APCDs collect information with sufficient identifying information to allow linkage of all claims for any given individual. They employ a centralized architecture, assembling the data rather than distributing queries as does Mini-Sentinel. Data in APCDs is collected under state public-health powers, subject to any applicable state law protections and state FOIA requests. Claims data from the federal Medicare program and the federal-state Medicaid program have typically not been included in APCDs. This is a major gap and APCDs are exploring with the federal government what DUAs should require if these data are to be included in APCDs.

Like cancer registry data, APCD data is subject to any applicable state law protections and any state law FOIA requests. Some state rules specify some exemptions from state FOIA requests for APCD data.[114] If information from these databases is released at too granular a level, they raise significant patient confidentiality concerns, however. No APCDs to date have released provider-specific information, but they may be expected to do so if they are to be used by the public to make cost or quality comparisons. State FOIA requests can be anticipated for such provider-specific data, illustrating the need for adequate confidentiality protection for patients.

INCREASING CONFIDENTIALITY PROTECTIONS

The federal and state programs described in this chapter present a variety of approaches important to protecting confidentiality when large-scale data sets are used for surveillance. Methods include data architectures that feature distributed query methods to avoid centralized data collection. They also include minimization of the extent to which identifying data are collected. When identifying data are collected, federal law provides more robust confidentiality protection than do the laws of some states. These federal protections do not apply to de-identified data; although some attempts do exist to protect these data through DUAs, supervision of compliance with these contractual agreements is limited.

Although some surveillance activities are possible with the use of highly abstract aggregate data (for example, the number of reports by state of a symptom such as diarrhea), or with data that have been de-identified to HIPAA standards, it is clear that the public-health surveillance activities occurring in the United States today are not limited to the use of de-identified data and would be less useful if they were so limited. HIPAA permits public-health authorities to acquire this data without individual authorization and as argued above there are a number of reasons why an individual authorization requirement would be problematic. Although important federal laws apply to protect data collected by federal agencies—the Privacy Act, the Public Health Service Act, and CIPSEA—state law protections are more uneven. There are also important ways in which federal protections might be enhanced. This section presents several other methods to protect confidentiality while enabling data use for surveillance purposes.

Increased Use of Distributed Query Architectures

The Mini-Sentinel distributed query structure avoids creation of a centralized database. Data remain in the possession and control of provider organizations originally collecting them. In addition to avoiding the security issues presented by any data repository, this structure has further advantages. Data remain within the trust relationship where they were originally collected; providers collecting the data remain accountable for any uses and disclosures of them. Patients with concerns have direct links to these providers and can hold them accountable for their data. Providers have the ability to examine queries and to determine whether they raise confidentiality concerns for their patients. Under present law, an important limitation to this protection is that patient requests for an accounting of disclosures of protected health information under HIPAA need not include disclosures for treatment, payment, or health care operations or for national security purposes, although more general public-health purposes are included in this accounting requirement.[115] Many of the other surveillance systems described in this chapter do not use the distributed query architecture structure.

Application of the HIPAA Minimum Necessary Standard When Identifying Information Is Collected for Public-Health Purposes

Another feature of Mini-Sentinel is that it applies the HIPAA minimum necessary standard to collection of any identifying information. This standard requires consideration of those who need access to information and types of information collected.[116] At present, the HIPAA minimum necessary standard does not apply to information required to be collected by other law.[117] Minimum necessary does apply to information collected for public-health activities, but entities responding to requests are permitted to rely on the public-health authority's determination of the information necessary.[118] A description of New York City's syndromic surveillance practices states the need for careful application of the minimum necessary standard to protect privacy.[119] A practical limitation at present on the application of this standard is that methods have not yet been developed for separate management of types of information in medical records that may be especially sensitive, such as substance abuse treatment information.[120]

Recommendations Regarding DUAs

DUAs are contractual agreements. As such, their enforcement and oversight is a matter of private law. The authors of this chapter have found no published cases involving litigation for violation of a DUA. DUAs may be functioning well but there is no way to be certain. Oversight mechanisms for DUAs should be augmented, including the requirements that these agreements be registered, publicly available, and subject to impartial third-party review and monitoring. One suggestion is the creation of special privacy boards for public-health data uses.[121]

Clarification of FOIA Exceptions

FOIA is a "sunshine" law expressing firm and well-justified commitment to open government. On the other hand, there may be a need to clarify the FOIA exceptions applicable to information collected for public-health surveillance. The FOIA exception for information protected by statute requires clear congressional intent that the information be confidential and, after 2009, specific citation to this FOIA provision. It is not clear that all of the federal surveillance activities described in this chapter would meet this requirement, especially as certificates of confidentiality and assurances of confidentiality have not been definitively tested in court. The FOIA exception for medical records is construed to allow courts to balance the importance of the privacy interest against public interests in having information available; according explicit statutory protection to specified data-collection activities would strike this balance as a matter of legislation instead of on a case-by-case basis. Moreover, surveillance activities conducted by the states are not subject to the federal FOIA statute.

Minimum State Law Standards

A significant problem for current surveillance is the variability of state data-protection laws. It would of course pose difficult problems for federalism to mandate a national FOIA standard for all state governments to meet. Under the structure of U.S. federalism, primary responsibility for protection of the public welfare, including public health, rests with the states.[122] The federal government has the power to regulate activities affecting health only through enumerated powers, such as the power to regulate interstate commerce. One possibility within this limit would be to circumscribe the exception to the HIPAA authorization requirement for disclosures required by public-health laws with a provision that authorization is required unless these laws accord at least the protections of the federal Privacy Act and FOIA protections. This would require patient authorization for disclosures unless state surveillance activities are public and unless state FOIA exceptions are at least as protective as the federal ones.

Augmented Publicity for Data Collection

At present, public-health surveillance activities are not well known—although the thought of "Big Brother" watching has become part of public rhetoric of concern. Federal Privacy Act notices may be buried in the Federal Register. Many states do not require publicity concerning surveillance activities. As Texas learned to great public-health loss of newborn screening samples, the price of apparent concealment may be very high. The public should be proactively informed and educated on an ongoing basis about surveillance activities and their risks and benefits.

Beyond the De-identification Divide

Neither the federal Privacy Act nor HIPAA apply to data lacking identifiers. Yet even setting risks of re-identification aside, such information may permit inferences about group characteristics that are harmful or stigmatizing and thus about individuals who are otherwise known to be part of the group. De-identified data may also be used in ways that some people find seriously objectionable. The distinction between data drawn from individual clinical records that continues to identify persons and data drawn from individual records that no longer identifies them is aimed at risks associated with identification but not at these other concerns. De-identified data thus should not be viewed as entirely beyond the purview of legal protection. At a minimum, these data-collection activities should be subject to publicity requirements so that they can be known and scrutinized. DUAs should be required for de-identified data to protect against inappropriate efforts at re-identification or data uses, subject to the oversight described above.

CONCLUSION

Surveillance serves many important public purposes. Reported data indicate that state epidemiologists encounter HIPAA barriers to syndromic surveillance.[123] The experience of Texas and Minnesota with public outcries against retention of newborn screening samples—an important resource for public health—also displays the potential for public concern. If the use of large-scale data sets for surveillance, adverse-event detection, and other important public-health purposes is to be placed on a secure basis, further and more consistent legal protections are imperative.

NOTES

1. For an overview of the benefits and issues involved in the use of large-scale sets of patient data for surveillance, together with an argument for public ownership of these data, see Marc A. Rodwin, *Patient Data: Property, Privacy & the Public Interest*, 36 Am. J.L. & Med. 586 (2010).

2. In such a "sectoral" approach to information privacy and confidentiality, the United States is a comparative outlier. The European Union, for example, has had a comprehensive directive for the protection of personal information for many years. *Directive 95/46/EC*, http://eur-lex.europa.eu/LexUriServ/LexUriServ.do?uri=CELEX:31995L0046:en:HTML.

3. 45 C.F.R. § 164.5; Khaled El Emam et al., *A Secure Protocol for Protecting the Identity of Providers When Disclosing Data for Disease Surveillance*, 18 J. Am. Med. Informatics Ass'n 212 (2011) describes a method for ensuring provider confidentiality while de-identifying data. This article attempts to respond to the concern that providers underreport. It develops an architecture to mask the identity of providers and to meet other requirements that are desirable for syndromic surveillance. It does preserve the ability to identify practices with unusual spikes to allow for public-health recontact.

4. 45 C.F.R. § 164.506 (2012).

5. *Id.* § 164.512(b) (2012).

6. Texas Department of State Health Services, Statement: Newborn Screening Settlement (December 22, 2009), http://www.dshs.state.tx.us/news/releases/20091222.shtm. Texas now is required to offer parents the option to request that their child's screening sample be destroyed, Texas Department of State Health Services, Use and Storage of Newborn Screening Bloodspot Cards, *available at* www.dshs.state.tx.us/lab/nbsdestructiondirective.pdf; Minnesota Department of Health, News Release, Minnesota Department of Health to begin destroying newborn blood spots in order to comply with recent Minnesota Supreme Court ruling (January 31, 2012), http://www.health.state.mn.us/news/pressrel/2012/newborn013112.html.

7. Nicholas P. Terry & Leslie P. Francis, *Ensuring the Privacy and Confidentiality of Electronic Health Records*, 2007 U. Ill. L. Rev. 701 (2007). In addition to information, "privacy" has also been used to conceptualize rights of individual choice over personal matters such as reproduction or rights to be free of intrusions into personal space such as by police searches.

8. The "Privacy Rule" of HIPAA, 45 C.F.R. pt. 164, is thus a misnomer; the rule is really a confidentiality rule.

9. Leslie P. Francis et al., *Syndromic Surveillance and Patients as Victims and Vectors*, 6 J. Bioethical Inquiry 187 (2009).

10. This is the language of the Health Insurance Portability and Accountability Act (HIPAA) Privacy Rule. 45 C.F.R. § 164.514(b)(1)(i).

11. *See, e.g.*, Khaled El Emam et al., *Evaluating Common De-Identification Heuristics for Personal Health Information*, J. Med. Internet Research, 2006; 8(4):e28, doi:10.2196/jmir.8.4.e28; Bradley Malin, Kathleen Benitez & Daniel Masys, *Never Too Old for Anonymity: A Statistical Standard for Demographic Data Sharing via the HIPAA Privacy Rule*, 18 J. Am. Med. Informatics Ass'n 3 (2011); Paul Ohm, *Broken Promises of Privacy: Responding to the Surprising Failure of Anonymization*, 57 UCLA L. Rev. 1701 (2010).

12. Staal Vinterbo, *Are There Limits to Privacy Preserving Sharing of Data?*, http://www.ncvhs.hhs.gov/110512p1.pdf (last visited May 14, 2011).

13. S.B. Trinidad at al., *Research Practice and Participant Preferences: The Growing Gulf*, 331 Sci. 287–88 (2011); Mark A. Rothstein, *Is Deidentification Sufficient to Protect Health Privacy in Research?*, 10 Am. J. Bioethics 3 (2010).

14. Leslie P. Francis & John G. Francis, *Group Compromise: Perfect Cases Make Problematic Generalizations*, 10 Am. J. Bioethics 25 (2010); Joan McGregor, *Racial, Ethnic, and Tribal Classifications in Biomedical Research with Biological and Group Harm*, 10 Am. J. Bioethics 23 (2010). The Havasupai won a significant settlement in litigation against Arizona State University. Amy Harmon, *Indian Tribe Wins Fight to Limit Research of Its DNA*, N.Y. Times, Apr. 21, 2010.

15. *See, e.g.*, Henry Rolka & Jean O'Connor, *Real-Time Public Health Biosurveillance: Systems and Policy Considerations*, *in* Infectious Disease Informatics and Biosurveillance, Integrated Series in Information Systems 3–22 (D. Zeng et al. eds.), 27, DOI 10.1007/978-1-4419-6892-1_1.

16. For a general discussion of secondary uses of health information and the need for protections, see National Committee on Vital and Health Statistics, Report to the Secretary of HHS, Enhanced Protections for Uses of Health Data: A Stewardship Framework for "Secondary Uses" of Electronically Collected and Transmitted Health Data (Dec. 21, 2007), *available at* http://www.ncvhs.hhs.gov/071221lt.pdf.

17. Despite the frequent allegation that privacy matters less in the "Facebook" age, there is considerable evidence that people continue to value privacy, albeit possible expressed in different forms, as well as that they may misunderstand the extent to which privacy is protected over the Internet. *See, e.g.*, Chris Jay Hoofnagle et al., *How Different Are Young Adults from Older Adults When It Comes to Information Privacy Attitudes and Policies* (Apr. 14, 2010), http://ssrn.com/abstract=1589864.

18. Alan P. Rubel, *Justifying Public Health Surveillance: Basic Interests, Unreasonable Exercise, and Privacy* (Jan. 11, 2012), *available at* http://ssrn.com/abstract=1991958.

19. *See, e.g.*, David Pozen, *Deep Secrecy*, 62 Stan. L. Rev. 257 (2010).

20. HHS is conducting annual health "datapaloozas" showcasing innovative technologies using their data. Todd Park, *Health Datapalooza 2012: Throw Your Hat Into the Ring!*, The Health Care Blog (Mar. 12, 2012), http://thehealthcareblog.com/blog/2012/03/12/health-datapalooza-2012-throw-your-hat-into-the-ring/. Todd Park, chief technology officer at HHS and a primary force behind "data liberation" at HHS, was appointed chief technology officer of the federal government in 2012.

21. Secretary's Advisory Committee on Automated Personal Data Systems, Records, Computers and the Rights of Citizens (July 1973), http://aspe.hhs.gov/DATACNCL/1973privacy/tocprefacemembers.htm,

22. 5 U.S.C. § 552a (2012).

23. *See, e.g.*, FAA v. Cooper, 2012 U.S. LEXIS 2539 (March 28, 2012).

24. 5 U.S.C. § 552a(b)(2), (8) (2012).

25. *Id.* § 552a(e)(6) (2012).

26. *Id.* § 552a(d)(1), (2) (2012).

27. *Id.* § 552a(e)(1) (2012).

28. *Id.* § 552a(e)(4) (2012). Thus the Privacy Act restricts the possibility of federal "deep secrets": possession of information when the public does not even know the information has been gathered. Secrets are "shallow," by contrast, when the public does not know the substance of the information but does know that it has been collected. *See* Pozen, *supra* note 19.

29. 5 U.S.C. § 552 (2012).

30. *Id.* § 552(a)(3) (2012).

31. *Id.* § 552(b)(3)(A) (2012).

32. *Id.* § 552(b)(3)(B) (2012).

33. *Id.* § 552(b)(4) (2012).

34. *Id.* § 552(b)(6) (2012).

35. *Id.* § 552(b) (2012).

36. RonNell Anderson Jones, *Litigation, Legislation, and Democracy in a Post-Newspaper America*, 68 Wash. & Lee L. Rev. 557 (2011); Testimony of Clark Hoyt, Representing Sunshine in Government Initiative, Freedom of Information Act, Hearing before the House of Representatives Committee on Oversight and Government Reform Subcommittee on Information Policy, Census, and National Archives (February 14, 2007).

37. U.S. Dep't of Justice, Freedom of Information Act Guide, May 2004, http://www.justice.gov/oip/reverse.htm (last visited Nov. 5, 2012).

38. Milner v. Dep't of the Navy, 131 S. Ct. 1259 (2011). The exemption is 5 U.S.C. § 552(b)(2) (2012). The records concerned data and maps used to help store explosives at a naval base.

39. An agency's personnel rules and practices were thus to be understood as concerning only "employee relations or human resources," and not as also encompassing materials that might be used by others to circumvent agency regulations. *Milner*, 131 S. Ct. at 1262.

40. Sorrell v. IMS Health Inc., 131 S. Ct. 2653 (2011) (Vermont statute restricting sales of provider-identifying prescription information violates constitutional protection of commercial speech); Brown v. Entm't Merchants Ass'n, 131 S. Ct. 2729 (2011) (California statute prohibiting the sale or rental of violent video games to minors without parental consent constitutionally overbroad).

41. Creed v.NTSB, 758 F. Supp. 2d 1 (D.D.C. 2010).

42. Creed v. NTSB, No. 10-1393, 2011 WL 1767839 (D.D.C. Apr. 7, 2011).

43. Highway Accident Report: Truck-Tractor Semitrailer Rear-End Collision Into Passenger Vehicles on Interstate 44 Near Miami, Oklahoma June 26, 2009 (Sept. 28, 2010), http://www.ntsb.gov/doclib/reports/2010/HAR1002.pdf (hereinafter "Driver's Medical Information").

44. The VICP was established in 1988 to replace personal injury litigation and thereby to encourage vaccine manufacturers to stay in the market. National Vaccine Injury Compensation Program, http://www.hrsa.gov/vaccinecompensation/index.html (last visited Nov. 5, 2012).

45. One set of advocates on behalf of parents seeking compensation for alleged vaccine-induced autism made a FOIA request to HHS for records concerning all decisions granting VICP compensation. When HHS responded that compliance would be immensely expensive and time-consuming, the requesters resorted to alternative methods of attempting to obtain the information rather than pressing the FOIA request. For a description of the FOIA request and

response, *see* Mary Holland et al., *Unanswered Questions from the Vaccine Injury Compensation Program: A Review of Compensated Cases of Vaccine-Induced Brain Injury*, 28 PACE ENVTL. L. REV. 480 (2011).

46. Long v. DOJ, 778 F. Supp. 2d 222 (N.D.N.Y. 2011).

47. Long v. DOJ, 703 F. Supp. 84 (N.D.N.Y. 2010).

48. The initial ruling had looked to the section of the Act requiring reporting of events listed in the vaccine injury table, 42 U.S.C. § 399aa-25, which specifically protected patient name, street address, telephone number, and medical records, but not locality of administration, name of the administering health care provider, date of vaccination, or information concerning the nature of the reported event.

49. This ruling relied on 42 U.S.C. § 300aa-12(d)(4)(A), providing that information submitted to a special master or a court in a petition for compensation may not be disclosed without consent to anyone not a party to the proceeding.

50. For a discussion of concerns that FOIA requests may be used for competitive advantage, see Sarah Taylor Roller et al., *FRA's Expanding Postmarket Authority to Monitor and Publicize Food and Consumer Health Product Risks: The Need for Procedural Safeguards to Reduct 'Transparency' Policy Harms in the Post-9/11 Regulatory Environment*, 64 FOOD & DRUG L.J. 577 (2009).

51. House v. Sec'y of HHS, 2011 U.S. Claims LEXIS 2549 (Fed. Cl., Feb. 28, 2011); Castagna v. Sec'y of HHS, 2011 U.S. Claims LEXIS (Fed. Cl. Spec. Mstr., Aug. 25, 2011). The section of the VICP Act relied on in these decisions is 42 U.S.C. § 300aa-12(d)(4)(B) (2012). The difference is that information submitted during the proceedings should receive more stringent protection than information relied on in the decision, which the public has an interest in knowing.

52. Douglas L. Sylvester & Sharon Lohr, *The Security of Our Secrets; A History of Privacy and Confidentiality in Law and Statistical Practice*, 83 U. DENV. L. REV. 147 (2005).

53. 13 U.S.C. § 9 (2012). There are limited exceptions not relevant to this discussion.

54. 42 U.S.C. § 242b (2012).

55. *Id.* § 242m(d) (2012). The exact statutory language reads:

> Information; publication restrictions. No information, if an establishment or person supplying the information or described in it is identifiable, obtained in the course of activities undertaken or supported under section 304, 306, or 307 [42 U.S.C.S. § 242b, 242k, or 242l] may be used for any purpose other than the purpose for which it was supplied unless such establishment or person has consented (as determined under regulations of the Secretary) to its use for such other purpose; and in the case of information obtained in the course of health statistical or epidemiological activities under section 304 or 306 [42 U.S.C.S. § 242b or 242k], such information may not be published or released in other form if the particular establishment or person supplying the information or described in it is identifiable unless such establishment or person has consented (as determined under regulations of the Secretary) to its publication or release in other form.

For a full description of the legal protections applying to the National Center for Health Statistics, see National Center for Health Statistics, NCHS Staff Manual on Confidentiality, *available at* http://www.cdc.gov/nchs/data/misc/staffmanual2004.pdf.

56. Akzo-Nobel Inc. v. United States, No. 00-30834, 2001 WL 34772206 (5th Cir. May 25, 2001). Note: the slip op. was provided by the CDC.

57. 44 U.S.C. § 3501.

58. *Id.* § 3504(g).

59. *Id.* § 3506(g).

60. Eleanor Singer, *Access to Research Data: Reconciling Risks and Benefits*, 15 J.L. & POL'Y 85 (2006).

61. CDC, *Certificates and Assurances of Confidentiality*, http://www.cdc.gov/od/science/integrity/confidentiality/.

62. 42 U.S.C. § 241(a), (d).

63. U.S. Department of Health and Human Services, National Institutes of Health Office of Extramural Research, *Certificates of Confidentiality: Background Information*, http://grants.nih.gov/grants/policy/coc/background.htm. The secretary has also delegated the authority to issue these certificates to the CDC. CDC, *Certificates and Assurances of Confidentiality*, http://www.cdc.gov/od/science/integrity/confidentiality/.

64. Health Resources and Services Administration, *Certificates of Confidentiality—Frequently Asked Questions*, http://www.hrsa.gov/publichealth/guidelines/HumanSubjects/faqs.html.

65. State v. Bradley, 634 S.E.2d 258 (N.C. App. 2006).

66. NIH Office of Intramural Research, *Certificates of Confidentiality Frequently Asked Questions*, http://grants.nih.gov/grants/policy/coc/faqs.htm (last visited June 20, 2011); Paul G. Stiles & John Petrila, *Research and Confidentiality: Legal Issues and Risk Management Strategies*, 17 PSYCHOL. PUB. POL'Y & L. 333, 333–56 (2011); Laura Beskow, Lauren Dame & E. Jane Costello, *Certificates of Confidentiality and Compelled Disclosure of Data*, 322 SCI. 1054, 1054–55 (2008).

67. *See* RonNell Anderson Jones, *Litigation, Legislation, and Democracy in a Post-Newspaper America*, 68 WASH. & LEE L. REV. 557 (2011).

68. Michael W. Field, *Rhode Island's Access to Public Records Act: An Application Gone Awry*, 8 ROGER WILLIAMS U. L. REV. 293 (2003).

69. Arkansas Dep't of Health v. Westark Christian Action Council, 910 S.W.2d 199 (Ark. 1995).

70. Kansas v. Comprehensive Health of Planned Parenthood of Kansas & Mid-Missouri, Inc., 241 P.2d 45 (Kan. 2010).

71. State *ex rel.* Dalton v. Mundy, 257 N.W.2d 877 (Wis. 1977).

72. Perkins v. Freedom of Info. Comm'n, 635 A.2d 783 (Conn. 1993).

73. S. Illinoisan v. Ill. Dep't of Health, 844 N.E.2d 1 (Ill. 2006).

74. Claire V. Broome, *Overview, Policy, and Systems: Federal Role in Early Detection Preparedness Systems*, Mortality & Morbidity Wkly. Rep. 54 (Suppl): 7–9 (Aug. 26, 2005).

75. Public Health Information Network, http://www.cdc.gov/phin/ (last visited Nov. 5, 2012).

76. *National Electronic Disease Surveillance System Working Group, National Electronic Disease Surveillance System (NEDSS): A Standards-Based Approach to Connect Public Health and Clinical Medicine*, 7 J. Pub. Health Mgmt. Prac. 43 (2001).

77. HITECH Act, Pub. L. No. 111-5. The meaningful-use program was authorized under the Health Information Technology for Economic and Clinical Health (HITECH) Act, as a way to improve health through the increased adoption and use of electronic records. The final rule implementing stage 1 of the meaningful-use incentive program is at 75 Fed. Reg. 44,314–44,588 (July 28, 2010).

78. 75 Fed. Reg. 44,370–44,375 (July 28, 2010).

79. Electronic submission of reportable lab results to public-health agencies is on the stage 1 menu set for eligible hospitals, Eligible Hospital Meaningful Use Table of Contents Core and Menu Set Objectives, https://www.cms.gov/EHRIncentivePrograms/Downloads/EP-MU-TOC.pdf. Hospitals must select five out of the ten listed items in the menu set. The Notice of Proposed Rule Making for Meaningful Use Stage 2 proposes to make the electronic submission of reportable lab results a core requirement for eligible hospitals for meaningful use stage 2, except where public-health departments do not have the capacity to receive these results. Department of Health and Human Services, Centers for Medicare and Medicaid Services, Medicare and Medicaid Programs; Electronic Health Record Incentive Program—Stage 2, http://www.ofr.gov/OFRUpload/OFRData/2012-04443_PI.pdf.

80. CDC, *State Electronic Disease Surveillance Systems—United States, 2007 and 2010*, 60 Morbidity & Mortality Wkly. Rep. 1421 (2011).

81. For example, the tuberculosis assurance is CDC, Online Tuberculosis Information System (OTIS) Technical Reference (Apr. 2011), http://wonder.cdc.gov/wonder/help/TB/OTISTechnicalReference.html#2.

82. George Hripcsak et al., *Syndromic Surveillance Using Ambulatory Electronic Health Records*, 16 J. Am. Med. Informatics Ass'n 354 (2009).

83. Richard Heffernan et al., System Descriptions New York City Syndromic Surveillance Systems, MMWR 53 (Suppl): 23–27 (Sept. 24, 2004), http://www.cdc.gov/mmwr/preview/mmwrhtml/su5301a6.htm (describing New York City's syndromic surveillance for waterborne illnesses established in 1995); Richard Heffernan et al., *Syndromic Surveillance in Public Health Practice, New York City*, Emerging Infectious Diseases [serial on the internet], http://wwwnc.cdc.gov/eid/article/10/5/03-0646.htm, DOI: 10.3201/eid1005.030646 (May 2004) (describing the introduction of the New York City Department of Health and Mental Hygiene's institution of syndromic surveillance in 2001 in the aftermath of the attacks on the World Trade Center). Interestingly, although New York's early experience indicated success in identifying outbreaks of infectious diseases such as influenza, norovirus, and rotavirus, it was not successfully in identifying outbreaks of food-borne illness.

84. C.C. van den Wijngaard et al., *Evaluation of Syndromic Surveillance in the Netherlands: Its Added Value and Recommendation for Implementation*, 16 Eurosurveillance (2011).

85. Nicholas E. Kmam & Daniel J. Bachmann, *Biosurveillance: A Review and Update*, Advances in Preventive Medicine (2012), doi:10.1155/2012/301408. Analyses that support the cost effectiveness of syndromic surveillance in the detection of bioterrorism are based on simulations. Kerry A. McBrien et al., *Use of Outcomes to Evaluate Surveillance Systems for Bioterrorist Attacks*, 10 BioMed Central Medical Informatics & Decision Making, doi: 10.1186/1472-6947-10-25 (May 7, 2010).

86. van den Wijngaard, *supra* note 84; Zalman Kaufman et al., *Evaluation of Syndromic Surveillance System Using the WSARE Algorithm for Early Detection of an Unusual, Localized Summer Outbreak of Influenza B: Implications for Bioterrorism Surveillance*, 9 Israeli Med. Ass'n J. 3 (2007). Even at local levels, however, sensitivity and specificity may be problematic; *see* K. Hope et al., *Syndromic Surveillance: Is It a Useful Tool for Local Outbreak Detection?*, 60 J. Epidemiology & Community Health 374 (2006).

87. BioSense Program Redesign, https://sites.google.com/site/biosenseredesign/about (last visited Nov. 5, 2012).

88. Public Health Security and Bioterrorism Preparedness and Response Act of 2002, Pub. L. No. 107-188, 116 Stat. 594, § 103.

89. Taha A. Kass-Hout et al., *CDC's BioSense2.0: Bringing Together the Science and Practice of Public Health Surveillance*, Am. J. Preventive Med. (Nov. 15, 2011), http://ajpmonline.wordpress.com/2011/11/15/cdc%E2%80%99s-biosense-2-0-bringing-together-the-science-and-practice-of-public-health-surveillance-4/; Colleen A. Bradley et al., *BioSense: Implementation of a National Early Event Detection and Situational Awareness System*, Mortality & Morbidity Wkly. Rep. 54 (Suppl.): 11–19 (Aug. 26, 2005).

90. James D. Nordin et al., *Bioterrorism Surveillance and Privacy: Intersection of HIPAA, the Common Rule, and Public Health Law*, 98 AM. J. PUB. HEALTH 802 (2008).

91. Public Health Information Network, *PHIN Messaging Guide for Syndromic Surveillance: Emergency Department and Urgent Care Data* (Oct. 2011), *available at* http://www.cdc.gov/phin/library/guides/PHIN_MSG_Guide_for_SS_ED_and_UC_Data_v1_0.pdf; Taha A. Kass-Hout & Charlie Ishikawa, *Summary of the CDC and ISDS Recommended Emergency Department and Urgent Care Core and Optional Data Elements for Syndromic Surveillance, available at* https://docs.google.com/viewer?a=v&pid=sites&srcid=ZGVmYXVsdGRvbWFpbnxiaW9zZW5zZXJlZGVzaWdufGd4OjY4ZDEwNjNiMzZmZTVkMTI&pli=1 (last visited Nov. 5, 2012).

92.

> We are employing industry-standard, state-of-the-art technologies and practices to ensure that only authorized users can access the system. We are also implementing access control tools for data providers, so that you can specify who within the community can access your data, and under what conditions. Lastly, as the system evolves, we will periodically engage outside security consultants to perform security audits and testing.

BIOSENSE PROGRAM REDESIGN, https://sites.google.com/site/biosenseredesign/about (last visited Nov. 5, 2012).

93. *PHIN Messaging Guide for Syndromic Surveillance: Emergency Department and Urgent Care Data, supra* note 91, at 9.

94. International Society for Disease Surveillance, *Final Recommendation: Core Processes and HER Requirements for Public Health Syndromic Surveillance* (Jan. 31, 2011), http://www.syndromic.org/uploads/files/ISDSRecommendation_FINAL.pdf.

95. Medicare and Medicaid Programs; Electronic Health Record Incentive Program, 42 C.F.R. Pts. 412, 413, & 495 (2012), *available at* http://federalregister.gov/a/2012-04443.

96. Department of Health and Human Services, *Centers for Medicare and Medicaid Services, Medicare and Medicaid Programs; Electronic Health Record Incentive Program—Stage 2*, http://www.ofr.gov/OFRUpload/OFRData/2012-04443_PI.pdf.

97. Blake Caldwell, Connecting for Biosurveillance: Essential BioSense Implementation Concepts, AMIA Blue Ribbon Panel (Apr. 27, 2006), http://www.amia.org/sites/amia.org/files/2006-Policy-Meeting-biosurveillance.pdf.

98. Margaret A. Dudeck et al., *National Healthcare Safety Network (NHSN) Report, Data Summary for 2010, Device-Associated Module*, 39 AM. J. INFECTION CONTROL 798 (2011); NATIONAL HEALTHCARE SAFETY NETWORK, www.cdc.gov/nhsn/ (last visited Nov. 5, 2012).

99. CDC, *National Healthcare Safety Network (NHSN)*, http://www.cdc.gov/nhsn/feature_dataStat.html.

100. CDC, *Data Use Agreement (DUA) Template*, http://www.cdc.gov/HAI/surveillance/DUA-announcment.html.

101. Under the DUA, states must

> specifically [agree] that, to the extent permitted by State and federal law, it will not release COVERED DATA requested under a State's open records laws; to media; for litigation purposes; that is proprietary and if disclosed could cause competitive harm; or to anyone other than department staff or contractors who are explicitly authorized to use those data for surveillance and/or prevention purposes only.

State-provided data must also apply appropriate security protections under the DUA.

102. Pub. L. No. 110-85, 121 Stat. 823, Title IX, sect. 905, 21 U.S.C. § 355(k)(3) (2012).

103. There is no effective surveillance for medical devices in the United States, although other countries keep registries. Frederic S. Resnic & Sharon-Lise T. Normand, *Postmarketing Surveillance of Medical Devices—Filling in the Gaps*, NEW ENG. J. MED., February 14, 2012 (10.1056/NEJMp1114865); R.G. Hauser, February 14, 2012, DOI: 10.1056/NEJMp1114695, "Here We Go Again—Another Failure of Postmarketing Device Surveillance"—example is "class I recall (a designation indicating a risk of serious injury or death) of the Riata and Riata ST implantable cardioverter-defibrillator (ICD) leads (St. Jude Medical), which are implanted in approximately 79,000 patients in the United States alone." There are organizations that maintain registries, for example the American College of Cardiology has a number of registries (member-only) for cardiac devices.

104. Mini-Sentinel, *Principles and Policies* (May 27, 2011), http://mini-sentinel.org/work_products/About_Us/Mini-Sentinel-Principles-and-Policies-May-2011.pdf.

105. MINI-SENTINEL, http://mini-sentinel.org/about_us/ (last visited Nov. 5, 2012); Interview by the author with Dr. Richard Platt, principal investigator of the Mini-Sentinel Program (March 4, 2012).

106. Mini-Sentinel, *Overview and Description of the Common Data Model v.2.1*, http://www.mini-sentinel.org/work_products/Data_Activities/Mini-Sentinel_Common-Data-Model_v2.1.pdf.

107. "Minimum necessary" is the standard the HIPAA Privacy Rule applies to the use of information for health care operations such as quality improvement activities, among other uses. 45 C.F.R. § 164.502(b). This is more caution than is legally required; neither the HIPAA Privacy Rule nor the HITECH Act's enhanced privacy protections would require applying the minimum necessary standard to public-health uses, however. HITECH Act § 13405(b)(3).

108. Sarah L. Cutrona et al., Mini-Sentinel Systematic Validation of Health Outcome of Interest: Acute Myocardial Infarction Cases Report (May 4, 2011), http://www.mini-sentinel.org/work_products/Validation_HealthOutcomes/Mini-Sentinel-Validation-of-AMI-Cases.pdf.

109. National Cancer Institute, *State Data Profiles*, http://statecancerprofiles.cancer.gov/.

110. 45 C.F.R. § 164.512(b).

111. S. Illinoisan v. Ill. Dep't of Public Health, 747 N.E.2d 401 (Ill. App. Ct. 2001).

112. ACPD Council, *Interactive State Report Map*, http://www.apcdcouncil.org/state/map.

113. For a description of APCDs, see Denise Love, William Custer & Patrick Miller, *All Payer Claims Databases: State Initiatives to Improve Health Care Transparency*, http://www.commonwealthfund.org/~/media/Files/Publications/Issue%20Brief/2010/Sep/1439_Love_allpayer_claims_databases_ib_v2.pdf.

114. APCD Council, *Step 4: Data Release Rules*, http://www.apcdcouncil.org/step-4-data-release-rules (last visited Nov. 5, 2012).

115. 45 C.F.R. § 164.528(a)(1)(i), (vi).

116. HHS, *Minimum Necessary Requirement*, http://www.hhs.gov/cor/privacy/hipaa/understanding/coveredentities/minimumnecessary.html.

117. 45 C.F.R. §§ 164.502(b)(2)(v), 164.512(a)(1).

118. 45 C.F.R. § 164.512(b); *Minimum Necessary Requirement*, *supra* note 116.

119. Richard Heffernan et al., *Syndromic Surveillance in Public Health Practice, New York City*, Emerging Infectious Diseases [serial on the internet], http://wwwnc.cdc.gov/eid/article/10/5/03-0646.htm, DOI: 10.3201/eid1005.030646 (May 2004).

120. Natl. Comm. on Vital & Health Stats., Letter to the Secretary, *Recommendations Regarding Sensitive Health Information*, http://www.ncvhs.hhs.gov/101110lt.pdf.

121. James D. Nordin et al., *Bioterrorism Surveillance and Privacy: Intersection of HIPAA, the Common Rule, and Public Health Law*, 98 Am. J. Pub. Health 802 (2008).

122. For a description of the relevance of federalism to the conduct of surveillance, see Rolka & O'Connor, *supra* note 15, at 11.

123. Daniel Drociuk, J. Gibson & J. Hodge, Jr., *Health Information Privacy and Syndromic Surveillance Systems*, MMWR Supplement 53 (Suppl.): 221–25 (Sept. 24, 2004).

Afterword

As Charlotte Hess and the late Nobel laureate Elinor Ostrom recently observed, biology is quickly becoming an information science.[1] As such, the legal issues that face those in the field of bioinformatics consist of one part biology, one part computer science, and one part arising from their unique combination.

Although bioinformatics has arguably been a scientific discipline for at least 20 years, the body of law surrounding it remains quite diffuse. Thus, we are left to break the field down into its component parts and address the body of law for each of them. Our intention in producing this book was to begin the assessment of bioinformatics law as a cohesive whole and to assemble in a single resource the laws, regulations, and considerations necessary for bioinformaticists and lawyers advising bioinformatics clients. Thus, this book highlights some of the unique legal issues that the field of bioinformatics will have to address in the years to come. Some of these issues affect any business or institution that makes substantial use of computer software, and others are unique to the intersection of biology and computing represented by bioinformatics.

Chapters 2, 3, and 4, which address issues relating to the licensing and development of computer software, highlight the unique challenges faced by bioinformaticists in this area. How does one decide whether it is preferable to develop a program in-house, download an open source code program from the Web, or license software from a commercial vendor? A range of issues must be considered, including price, availability, customizability, maintenance and support, liability, and potential risk. As a result, it is likely that every bioinformatics facility will include a combination of these different options, and it will take a careful lawyer to balance the requirements and obligations associated with this potentially volatile admixture.

Patents, of course, have been much in the news lately. Chapters 5 and 6 offer an overview of the patent issues affecting computer software and the biological entities that bioinformatics applications analyze, model, and represent. Many legal decisions hinge on a solid understanding of patent law, including whether to buy or build a software system, what sort of patent analysis to conduct prior to, or during, bioinformatics research, whether to apply for patents of one's own, and what protection and indemnification to seek from vendors of software and other bioinformatics systems, as well as research partners and collaborators and suppliers of biological materials. These chapters offer a detailed overview of the issues and risks associated with patents in this field. Chapter 7 offers a more cautionary view of biotechnology patents, and makes the reader question the implications of such patents for both personal privacy and social policy. We hope that the diversity of perspectives presented in these chapters gives the reader a flavor for the debates over patents that are currently ongoing in the courts and legislatures, as well as in the public arena.

Chapters 8 and 9 shift to two principal environments in which bioinformatics research is carried out: biotech-pharmaceutical collaborations and academic research institutions. Chapter 8 outlines the many commercial and legal considerations that must be considered when a biotechnology company wishes to explore a partnership with a larger pharmaceuticals vendor. Chapter 9 describes the university research environment in the United States, including modes in which bioinformatics research is conducted, legal aspects relating to patenting and the Bayh-Dole Act, and policy initiatives such as socially responsible licensing.

Chapter 10 addresses the emerging legal issues associated with developing technical standards for bioinformatics research. These include the avoidance of potentially anticompetitive practices during standard-setting, and implementing policies that will avert later disputes concerning patents and copyrights covering technical standards.

Chapters 11 and 12 address issues arising from the collection and storage of data, a hallmark of bioinformatics research. In Chapter 11, the author summarizes the legal status of database protection in the United States and Europe, pointing out differences and similarities and making recommendations regarding best practices for database protection. Chapter 12 covers data release requirements of major U.S. government (NIH) and international funding programs, including requirements that genomic and other biomedical data be deposited in publicly managed databases within prescribed time periods after generation.

Chapters 13, 14, and 15 focus on the personal implications of data collection and aggregation. In Chapter 13, the authors address the pressing information security and privacy challenges that bioinformatics applications invariably raise, as well as the expanding number of statutory and regulatory requirements arising in this area (HIPAA, GINA, HITECH, and a range of state law provisions). The authors also explore the implications of new and emerging technologies on data security and privacy, including the legal risks associated with the expanding use of mobile devices and the Internet for health care applications. Chapter 14 addresses the critical issue of patient informed consent in the context of DNA databanks, and extends the traditional notion of consent to new contexts such as data mining, population-level consent, and a charitable trust model for informed consent. In Chapter 15, the authors explore the emerging area of public-health surveillance and informatics, outlining both the ethical considerations and the many legal and regulatory regimes (the Privacy Act, FOIA, certificates of confidentiality, and state law regulations) that are implicated by these practices.

In 2001, shortly after the joint publication of the initial human genome sequence articles by the Human Genome Project (HGP) and Celera Genomics, Francis Collins, director of the HGP, urged the world to "cease using the phrase 'the post-genomic era,' at least for a decade or two."[2] Genomic science, Collins argued, was at a beginning, not an end. Though we have tried to be as comprehensive as possible, we are painfully aware that the field of bioinformatics is growing and evolving at a rate faster than any of us can truly appreciate. Thus, by the time this book goes to print, we are confident that new developments, both technological and legal, will have arisen, and that our advice, thoughts, and words will soon become obsolete. But we sincerely hope that, at least for a time, we have presented a good overview of the legal issues that affect the field of bioinformatics in these early days of the post-genomic era.

NOTES

1. Charlotte Hess & Elinor Ostrom, *A Framework for Analysing the Microbiological Commons*, 58 INT'L SOC. SCI. J. 335, 335 (2006).
2. Francis S. Collins, *Contemplating the End of the Beginning*, 11 GENOME RES. 641, 643 (2001).

About the Authors

Maria Bottis is an assistant professor in the Department of Archives and Library Science at Ionian University, Greece. Her main areas of interest are information law and ethics, and medical law and ethics. Professor Bottis holds LL.M. degrees from Cambridge University Law School and from Yale Law School. She also received an LL.B. (Rotary Award) and an honors Ph.D. in medical information law and ethics from Athens Law School.

Jordan J. Bowler is an associate at the law firm Davis Wright Tremaine. Mr. Bowler's practice focuses on intellectual property and other commercial transactions, in areas including business formation, contract, copyright, patent, trademark, data privacy, and employment law. He received a B.S. in applied mathematics and economics as well as a B.A. in political science from the University of Washington, and a J.D. from Stanford Law School.

Charley F. Brown is an associate in the intellectual property department of the law firm Ballard Spahr in Atlanta. His practice includes patent, copyright, trademark, and trade secret litigation; patent prosecution; client counseling; and opinion work. He is a registered patent attorney and conducted graduate research focused on genome and proteome annotation techniques, sequence analysis, structure prediction, genome assembly improvement, and microarray data analysis. Mr. Brown received a B.B.A. degree from Mercer University, an M.S. in bioinformatics from Georgia Institute of Technology, and his J.D. from Mercer University, Walter F. George School of Law.

Jeffrey S. Carter-Johnson is an adjunct professor at Michigan State University College of Law and the coordinator for the Law College's Intellectual Property Start-Up Project. His research interests include the regulation of university research and technology transfer from university research to the private sector. Dr. Carter-Johnson received a B.S. in biology at Old Dominion University. He then received his Ph.D. in microbiology at the University of Virginia, where his research was in the fields of virology and molecular biology. He received his J.D. from Washington University School of Law in St. Louis.

Jennifer Carter-Johnson is an associate professor of law at Michigan State University College of Law. Dr. Carter-Johnson focuses her research on intellectual property law and policy, investigating issues at the intersection of biological research and the law. She received a B.S. in mathematics and biology from Union University, a Ph.D. in microbiology from the University of Virginia, and a J.D. from the University of Michigan Law School.

Jorge L. Contreras is an associate professor of law at American University, Washington College of Law. Prior to joining American University, he served as acting director of the intellectual property program at Washington University in St. Louis, and was a partner at the law firm Wilmer Cutler Pickering Hale and Dorr in Boston, London, and Washington, D.C. Professor Contreras currently serves as co-chair of the National Conference of Lawyers and Scientists and co-chair of the ABA Section of Science & Technology Law Committee on Technical Standardization, and was the editor of the ABA's *Standards Development Patent Policy Manual* (2007). Professor Contreras is a member of the Advisory Council of NIH's National Center for Advancing Translational Science and formerly served on the Council of the National Human Genome Research Institute at NIH. He received his B.S.E.E. and B.A. (English) degrees from Rice University and his J.D. from Harvard Law School.

A. James Cuticchia recently formed AJC Legal Services, which focuses on providing services supporting entrepreneurship in the Research Triangle Park area in North Carolina. Additionally, he is an adjunct full-professor at the North Carolina Central University School of Law. He also teaches undergraduate courses in the BRITE program at NCCU that include Patent Law and FDA regulations. Dr. Cuticchia received a Ph.D. in genetics from the University of Georgia. He went on to perform genetic research in academic, not-for-profit, and biotechnology companies for 20 years before receiving a law degree at North Carolina Central University.

Jane Eckels is a partner at the law firm Davis Wright Tremaine, where she is deputy chair of the technology, e-business and digital media practice. Ms. Eckels practices in the areas of information technology and intellectual property, with a focus on e-commerce and information technology transactions. Ms. Eckels received an A.B. in French from Dartmouth College and a J.D. from the University of Michigan Law School.

John G. Francis is professor of political science at the University of Utah and was senior associate vice president for academic affairs at Utah from 1996 to 2011. For Hilary and Trinity terms, 2012, he was Beaufort Visiting Fellow (elected) at Lady Margaret Hall, University of Oxford. His research interests include citizenship and migration, federalism, and politics and public-health regulation.

Leslie P. Francis is a distinguished professor of law and philosophy and Alfred C. Emery professor of law at the University of Utah. She teaches and writes extensively in the areas of health law, bioethics, and disability. Dr. Francis received a B.A. from Wellesley College and a Ph.D. in philosophy from the University of Michigan. After joining Utah's philosophy faculty, Dr. Francis received her J.D. from the University of Utah.

Jon M. Garon is a professor of law and director of the Law + Informatics Institute at the Salmon P. Chase College of Law at Northern Kentucky University. His teaching and scholarship often focus on business innovation and the development of best business practices regarding the exploitation and effectiveness of the information and data systems in business, health care, media, and entertainment, and the public sector. Mr. Garon received a B.A. from the University of Minnesota and a J.D. from Columbia Law School.

Michele K. Herman is chief corporate counsel at Intellectual Ventures. Prior to joining Intellectual Ventures, Ms. Herman was a partner at Davis Wright Tremaine, and prior to that, associate general counsel at Microsoft Corporation, where her practice focused on technology standards, open source software, and technology and patent licensing. Ms. Herman has held leadership roles in open source and standard committees in several professional organizations. She is also an adjunct professor at Seattle University School of Law. Ms. Herman received her B.S. in electrical engineering and her J.D. from Rutgers University and is registered to practice before the U.S. Patent and Trademark Office.

Gregory J. Kirsch is a partner and registered patent attorney in the law firm of Smith, Gambrell & Russell, LLP, in its intellectual property practice group. He serves as patent counsel to clients ranging from large multinational corporations to small start-ups, as well as universities. His practice encompasses the entire range of patent and IP law, including patent procurement, strategic patent portfolio development, patent opinions (infringement, validity, and patentability), patent reissues and reexaminations, IP due diligence, licensing, and patent litigation. In addition to his full-time legal practice, Mr. Kirsch serves as an adjunct professor of patent and IP law at both Emory University School of Law in Atlanta and Washington University School of Law in St. Louis. He received a B.S. in electrical and computer engineering from the University of Iowa and a J.D. from Washington University School of Law.

Jeremy T. Marr is a member of the technology transactions and licensing group at Wilmer Cutler Pickering Hale and Dorr in Boston. His law practice focuses on life sciences and other high-technology transactional matters. He previously held a wide range of positions at various start-up, private, and public software companies, primarily focused on tools that increase the software development productivity of information technology teams. Mr. Marr received his undergraduate degree in computer science and psychology from Dartmouth College and his J.D. from Boston College Law School, where he is also an adjunct professor.

Heather J. Meeker is the chair of the IP/IT licensing and transactions group at the law firm Greenberg Traurig. Her practice focuses on intellectual property transactions for technology clients, covering a range of industries including software, communications, educational testing, computer equipment, and medical devices. Ms. Meeker holds a B.A. in economics from Yale University and a J.D. from the University of California at Berkeley School of Law.

Joanne Montague is an associate at the law firm Davis Wright Tremaine. Ms. Montague's practice focuses on intellectual property law, emphasizing open source software and standards as well as technology licensing. Ms. Montague received a B.S. in electrical engineering from Stanford University, an M.B.A. from the University of Pennsylvania, and a J.D. from Seattle University School of Law.

Arthur E. Peabody Jr. is lead Medicare counsel at the BlueCross BlueShield Association, where he heads the nationwide, multimillion-dollar acute care, Medicare Part A reimbursement appeals program. As an assistant United States attorney for the Eastern District of Virginia, he litigated civil cases as both plaintiff and defendant, including health care fraud actions. For many years, Mr. Peabody served as chief of the special litigation section in the civil rights division of the U.S. Department of Justice, where he managed all aspects of investigations and litigation to vindicate the federal rights of juvenile, elderly, and disabled persons around the country. He earned his J.D. degree from Cornell Law School.

Richard A. Spinello is an associate research professor in the Carroll School of Management at Boston College, where he teaches courses on ethics, social issues in management, and corporate strategy. Dr. Spinello has written and edited nine books on applied ethics and related areas, including *CyberEthics: Morality and Law in Cyberspace* and *A Defense of Intellectual Property Rights*. Dr. Spinello received an M.B.A. from the Carroll School of Management at Boston College and a Ph.D. from Fordham University.

Herman T. Tavani is professor emeritus of philosophy at Rivier University, a visiting scholar at the Harvard School of Public Health, and a past president of the International Society for Ethics and Information Technology. Dr. Tavani has written, edited, or coedited five books on social and ethical aspects of information technology. He holds B.A. and M.A. degrees from West Chester University and a Ph.D. from Temple University.

Lucy L. Thomson is principal of Livingston PLLC, a law firm that focuses on legal issues related to new technologies, information security, global data privacy, and compliance. She is chair of the ABA Section of Science & Technology Law. Appointed consumer privacy ombudsman in 14 federal bankruptcy cases, she has overseen the disposition of 200 million electronic consumer records. A former federal criminal prosecutor, Ms. Thomson worked for ten years as a senior principal engineer (information security) and privacy advocate at CSC, a global technology company. Editor of the *Data Breach and Encryption Handbook* (ABA 2011), she organized and spoke on panels at the RSA conferences in San Francisco and London. Ms. Thomson earned an M.S. degree from Rensselaer Polytechnic Institute in 2001, and holds a J.D. degree from the Georgetown University Law Center.

Kimberly J. Wade is a counsel in the technology transactions and licensing practice group at the law firm Wilmer Cutler Pickering Hale and Dorr LLP. Ms. Wade's practice focuses on intellectual property and technology licensing and transactions. She has represented biotechnology and information technology clients in transactions involving the development, licensing, manufacture, supply, and commercialization of products and technology. Ms. Wade holds a B.A. from Howard University and a J.D. from the University of Pennsylvania Law School.

Index